Malte Schemmann

Biaxial Characterization and Mean-field Based Damage Modeling of Sheet Molding Compound Composites

Schriftenreihe
Kontinuumsmechanik im Maschinenbau
Band 13

Karlsruher Institut für Technologie (KIT)
Institut für Technische Mechanik
Bereich Kontinuumsmechanik

Hrsg. Prof. Dr.-Ing. habil. Thomas Böhlke

Eine Übersicht aller bisher in dieser Schriftenreihe erschienenen Bände finden Sie am Ende des Buchs.

Biaxial Characterization and Mean-field Based Damage Modeling of Sheet Molding Compound Composites

by
Malte Schemmann

KIT Scientific Publishing

Dissertation, Karlsruher Institut für Technologie
KIT-Fakultät für Maschinenbau

Tag der mündlichen Prüfung: 16. Mai 2018
Referenten: Prof. Dr.-Ing. Thomas Böhlke
Prof. Dr. Andrew N. Hrymak
Prof. Dr.-Ing. Thomas Seelig

Impressum

KIT Scientific Publishing

Karlsruher Institut für Technologie (KIT)
KIT Scientific Publishing
Straße am Forum 2
D-76131 Karlsruhe

KIT Scientific Publishing is a registered trademark
of Karlsruhe Institute of Technology.
Reprint using the book cover is not allowed.

www.ksp.kit.edu

Print on Demand 2018 – Gedruckt auf FSC-zertifiziertem Papier

ISSN 2192-693X
ISBN 978-3-7315-0818-2
DOI 10.5445/KSP/1000084270

Biaxial Characterization and Mean-field Based Damage Modeling of Sheet Molding Compound Composites

Zur Erlangung des akademischen Grades

Doktor der Ingenieurwissenschaften

der Fakultät für Maschinenbau

Karlsruher Institut für Technologie (KIT)

genehmigte

Dissertation

von

MSc Malte Schemmann

Tag der mündlichen Prüfung:	16.05.2018
Hauptreferent:	Prof. Dr.-Ing. Thomas Böhlke
Korreferent:	Prof. Dr. Andrew N. Hrymak
Korreferent:	Prof. Dr.-Ing. Thomas Seelig

Zusammenfassung

Aufgrund ihrer hohen Steifigkeit und Festigkeit werden faserverstärkte Polymere (FRP) vermehrt im Automobilleichtbau eingesetzt. Vorteile von diskontinuierlich faserverstärkten Polymeren (DiCoFRP) sind insbesondere die Designfreiheit und die sehr wirtschaftliche Großserienfertigbarkeit. Kontinuierlich faserverstärkte Polymere (CoFRP) hingegen zeichnen sich durch ihre herausragende Steifigkeit und Festigkeit aus. Die hier beschriebenen Forschungsergebnisse entstanden im Rahmen des deutsch-kanadischen DFG Graduiertenkollegs GRK 2078, welches sich mit einem ganzheitlichen Ansatz für diskontinuierlich faserverstärkte Polymere mit kontinuierlichen Verstärkungen (CoDiCoFRP) befasst. Das Ziel bei der Verwendung dieser neuen Materialklasse besteht darin, DiCoFRP und CoFRP so zu kombinieren, dass die Vorteile beider Materialien vereint werden.

In dieser Arbeit liegt der Fokus auf DiCoFRP, speziell auf Sheet Molding Compounds (SMC) basierend auf ungesättigtem Polyurethan Polyester Hybridharz (UPPH), auf Epoxy und auf Vinylester (VE) Harzsystemen. Die inhomogene, anisotrope Faserorientierungsverteilung bestimmt die mechanischen Eigenschaften von SMC. Ein Kernelement dieser Arbeit ist die mikrostrukturabhängige Charakterisierung und Modellierung von SMC.

Der heterogene Spannungs- und Dehnungszustand einer biaxial belasteten Kreuzprobe erlaubt es im Allgemeinen nicht, die Materialparameter direkt über die Konstitutivgleichungen zu bestimmen. Deshalb wird ein mikrostrukturbasiertes Schema zur Parameteridentifikation von SMC mit inhomogener Faserorientierungsverteilung eingeführt.

Vergleiche zwischen dem, auf inhomogenen Materialparametern basierenden, simulierten Dehnungsfeld und dem gemessenen Dehnungsfeld werden im Detail diskutiert.

Eine große Herausforderung beim Design von biaxialen Kreuzproben liegt darin, einen hohen Dehnungszustand in der Probenmitte zu erhalten und gleichzeitig das Versagen in den eingespannten Probenarmen zu vermeiden. Es werden unterschiedliche Designs von Kreuzproben und zwei Konzepte zur Verstärkung der Probenarme vorgestellt. Die Untersuchung beinhaltet die beiden Lastszenarien, uniaxialer Zug und äquibiaxialer Zug. Bezüglich der eingeführten Optimalitätskriterien erweist sich ein Probentyp als optimal, bei dem auf der Ober- und Unterseite des SMC unidirektional verstärkte Schichten (CoFRP) aufgebracht wurden. Ein schonender Fräsprozess legte das reine SMC in der Probenmitte frei.

Zudem wird ein mikromechanisches Schädigungsmodell eingeführt. Das Modell bildet die dominierenden Schädigungsmechanismen – Matrix-Schädigung und Faser-Matrix Interface-Ablösung – im Rahmen eines Mori-Tanaka Homogenisierungsschemas ab. Die Matrix-Schädigung wird als phasengemittelte, isotrope Steifigkeitsdegradation modelliert. Die Interface-Schädigung wird durch eine äquivalente Interface-Vergleichsspannung auf der Fasermantelfläche bestimmt. Diese äquivalente Interfacespannung berücksichtigt die inhomogene Spannungsverteilung am Faser-Matrix Interface. Die Simulationsergebnisse geben Einblick in die Schädigungsentwicklung, sowie den daraus resultierenden Anteilen lasttragender Fasern bei unterschiedlichen Lastszenarien.

Summary

Due to their high specific strength and stiffness, fiber-reinforced polymers (FRP) are increasingly applied in lightweight components in the automotive industry. Advantages of discontinuous fiber-reinforced polymers (DiCoFRP) include good formability and economical mass production. Continuous fiber-reinforced polymers (CoFRP) offer outstanding stiffness and strength. This research was conducted within the German-Canadian research training group DFG GRK 2078, which focuses on integrated engineering of discontinuous fiber-reinforced polymers with continuous fiber reinforcements (CoDiCoFRP). The goal of this new material class is to combine CoFRP and CiCoFRP in such a way that the advantages of both material classes are maintained.

The herein presented thesis considers DiCoFRP, specifically an unsaturated polyester polyurethane hybrid resin (UPPH), vinylester (VE), and an epoxy resin sheet molding compound (SMC). The inhomogeneous and anisotropic fiber orientation distribution determines the mechanical behavior of the SMC. This work aims to characterize and model the SMC based on its microstructure.

For the heterogeneous stress and strain state under biaxial loadings of cruciform specimen it is, in general, not possible to analytically identify the material parameters of a constitutive equation. A microstructure-based parameter identification scheme for the SMC with an inhomogeneous fiber orientation distribution is introduced. The strain field simulated with the obtained inhomogeneous material parameters and the measured strain field on the biaxial specimen are compared.

A key challenge in the design of a cruciform biaxial tensile specimen lies in the design of a cruciform specimen that achieves a high strain level in the center region and prevents premature failure in the clamped specimen arms. Different cruciform specimen designs, including two concepts to reinforce the specimens' arms are introduced. An experimental evaluation includes two different loading scenarios, uniaxial tension and equi-biaxial tension. The best fit in terms of the defined optimality criteria is a specimen manufactured in a layup with unidirectional reinforcing outer layers. A gentle milling process exposed the pure SMC in the center region of the specimen.

Additionally, a micromechanical mean-field damage model for the SMC is introduced. The model captures the dominant damage mechanisms – matrix damage and fiber-matrix interface debonding – within a Mori-Tanaka homogenization framework. The matrix damage is modeled as phase-averaged isotropic stiffness degradation. The interface damage is governed by an equivalent interface stress on the lateral fiber surface. The inhomogeneous stress distribution in the fiber-matrix interface is taken into account in the definition of the equivalent stress governing the fiber-matrix interface damage. The simulation results provide an insight into the damage evolution for different loading scenarios.

Acknowledgments

Special thanks go to my supervisor Thomas Böhlke, who awakened my interest in mechanics and provided the continuum mechanics tools. His advice and critical discussions have made me the researcher and teacher I am today.

I thank my co-supervisor Andy Hrymak for our fruitful discussions, which genuinely broadened my perspective on composites and damage modeling. I am grateful for Andy's invitation to a research stay at the University of Western Ontario and the Fraunhofer Project Center @ Western. Andy has the unique ability to be able to find the few words that have had a huge positive impact on me.

I acknowledge Thomas Seelig for the engagement as my co-supervisor. He dedicated countless hours of his time on detailed discussions on damage modeling as well as to manuscript editing. His mechanical understanding of composites was a true enrichment to this work.

I thank Sven Matthiesen for chairing the doctoral committee and the related organizational aspects regarding this doctoral thesis.

I am thankful for all the collaborative discussions with my colleagues from the international research training group GRK 2078. Special thanks go to Pascal Pinter (IAM-WK), who provided the micro-computed tomography images within this thesis, David Bücheler (Fraunhofer ICT), who manufactured the composite plates, Loredana Kehrer (ITM-CM) for manufacturing the neat resin samples, Anton Helfrich (wbk) for milling the cruciform specimen, Michael Schober (Fraunhofer IWM) for an REM image, and Anna Trauth (IAM-WK) for the macroscopic characterization and extensive effort in manufacturing the neat resin samples.

My stay at the Fraunhofer Project Center @ Western would not have been nearly so enjoyable had it not been for Vanja, Dan, Louis, Sebastian, Antje, Kay, and Nils. Additionally, I would like to thank Jeff Wood for his organizational support and collaboration on a student project.

I thank Ute and Helga for all their administrative help throughout the last years. Thank you, Tom for your open ear regarding any organizational issues, Linux problems, and questions concerning engineering mechanics.

This thesis would not have been possible without the work of my students Sebastian, Thomas, Julian, Lukas, and Kevin. Your input and fresh point of views were a true enrichment.

My roommates Eric, Vedran, Johannes, Juliane, and Julian have helped guide me and our interactions have lead to great discussions about mechanics and many other topics.

I thank all of my colleagues at the ITM-CM for their support, nice coffee breaks, and just a great time overall.

I thank my family, especially Elke, Heinrich, and Steffen for their understanding support in any situation. Franzi, your wholehearted support and encouragement were extremely important to me. Matze, Stefan, Alex, Andy, and an awesome volleyball team, on and off the court, you have kept me motivated and provided me with a much-needed distraction.

The research documented in this doctoral thesis has been funded by the German Research Foundation (DFG) within the International Research Training Group "Integrated engineering of continuous-discontinuous long fiber reinforced polymer structures" (GRK 2078). The support by the German Research Foundation (DFG) is gratefully acknowledged.

Karlsruhe, May 2018 Malte Schemmann

Contents

Chapter 1

Introduction

1.1 Motivation

The global drive to reduce carbon dioxide emissions constitutes a key motivation for lowering fuel consumption in the automotive sector. Here, lightweight design plays a significant role, since a 100kg weight savings reduces fuel consumption by an average of 0.40l/100km for cars, and 0.49l/100km for light trucks (Bandivadekar et al., 2008). In the literature, the correlation between 100kg weight reduction and fuel savings varies from 0.15l/100km to 0.7l/100km (Bandivadekar et al., 2008). Economically, the automotive industry is willing to pay US$1.20 to US$13.70 per kilogram of weight savings (Bandivadekar et al., 2008). Heuss et al. (2012) predicted that, by 2030, automotive OEMs will pay up to €20 per kilogram of weight savings.

One weight-reduction strategy is to use lighter construction materials. The outstanding mass-specific strength and stiffness of fiber-reinforced polymers (FPR) have drawn considerable attention to this class of composites. FRP are sub-divided into continuous fiber-reinforced polymers and discontinuous fiber-reinforced polymers.

Continuous fiber-reinforced polymers (CoFRP), also known as endless fiber-reinforced polymers, consist of a polymer matrix and aligned fibers similar in length to the part dimensions. CoFRP permit fiber content of up to 60vol.%. The high fiber content and fiber alignment result in high stiffness and strength in the fiber direction(s). Moreover, the known fiber direction(s) allow for straight-forward modeling and dimensioning of CoFRP. Drawbacks include high cycle times, high scrap rates, and extensive trimming.

Typical manufacturing processes for discontinuous fiber-reinforced polymers (DiCoFRP) are injection and compression molding of a fiber-matrix suspension. This moldfilling behavior limits the fiber volume fraction, since the suspension requires a low viscosity, but it also allows for great design freedom, including the use or ribs, for example. Injection and compression molding permit low cycle times and thus promote economic application in high-volume production. The stiffness and strength of DiCoFRP are significantly lower than those in CoFRP. Although DiCoFRP is already applied in non-structural automotive components (Ernst et al., 2006), its use in structural components is hindered by a lack of robust and efficient dimensioning and characterization methods. One main challenge in modeling DiCoFRP is the process-induced heterogeneous fiber orientation distribution that is not known a priori.

A new hybrid class of materials is the so-called discontinuous fiber-reinforced polymer with continuous fiber reinforcements (CoDiCoFRP). The goal of this new material class is to combine CoFRP and DiCoFRP in such a way that the advantages of both material classes are maintained: high stiffness and strength along load paths, good formability, and low cycle times. Figure 1.1 shows the classification of composites with respect to fiber length. Even in the academic community, only preliminary investigations of this material class exist. FRP is also classified according to its matrix material. Fiber-reinforced thermosets (FRTS) have a thermoset matrix, whereas fiber-reinforced thermoplastics (FRTP) have a thermoplastic matrix material. The abbreviations CoFRTS, DiCoFRTS, CoDiCoFRTS, CoFRTP, DiCoFRTP, and CoDiCoFRTP are used analogously. Figure 1.2 shows an example of a CoDiCoFRTS component. The part consists of glass fiber DiCoFRTS (light regions) in and carbon fiber CoFRTS (black regions).

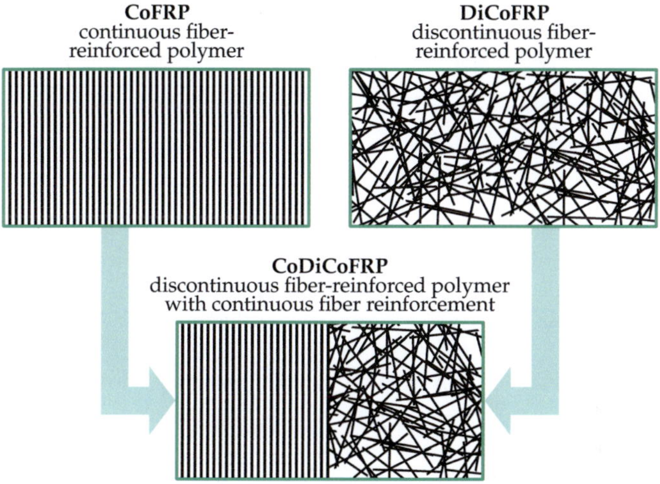

Figure 1.1: Classification of different fiber-reinforced polymer classes with respect to the fiber orientation

Figure 1.2: GRK 2078 reference structure made of discontinuous fiber-reinforced thermoset with continuous fiber reinforcements (manufactured at Fraunhofer Institute of Chemical Technology (ICT) Pfinztal)

Manufacturing technologies and materials for the introduced FRP classes include (Henning and Moeller, 2011):

- CoFRTP: tape laying, pultrusion
- CoFRTS: most traditional lamination processes, handlamination, resin transfer molding (RTM), pultrusion, filament winding
- DiCoFRTP: long-fiber-reinforced thermoplastic (LFT) compression and injection molding, short-fiber thermoplastic injection molding and foaming
- DiCoFRTS: sheet molding compound (SMC), bulk molding compound (BMC), reactive injection molding, fiber spraying.

This thesis presents research results from project S2 in the international research training group GRK 2078. The German-Canadian research program GRK 2078 focuses on an integrated engineering approach for CoDiCoFRP parts. The program's goal is to enable use of CoDiCoFRP in high-volume structural components by providing the necessary manufacturing technology, the methods for characterization and simulation, and the optimization tools. GRK 2078 consists of four research areas: characterization, simulation, technology, and design. The project S2 belongs to the simulation research area. The aim of research area simulation is to meet a foreseeable industry demand for a precise and robust virtual process chain that can shorten development cycles and reduce or even eliminate costly prototyping.

One contribution of project S2 is a mean-field model for two-scale structural finite element simulations within the virtual process chain. This thesis focuses on DiCoFRTS by means of sheet molding compound (SMC). A mechanism-based model of the damage behavior is presented. A key challenge lies in the process-dependent, heterogeneous fiber orientation distribution, which in turn, causes heterogeneous anisotropic material behavior. The fiber orientation distribution – one input to the microstructure-based mean-field model – was obtained by micro-computed tomography or moldfilling simulations.

A second goal of this thesis is to characterize DiCoFRTS under biaxial stress states. Here, first results on the parameter identification of materials with inhomogeneous microstructures are presented and an optimized specimen design is suggested. The anisotropic damage model is validated with biaxial tensile tests. The characterization of DiCoFRTS under biaxial stress states covers a wide range of application load cases for typical shell-like SMC structures.

1.2 State of the art[1]

1.2.1 Biaxial characterization and parameter identification of DiCoFRP

The biaxial characterization of DiCoFRP is still in the early stages of research. In this section, first a literature overview of biaxial tensile testing is provided, and then the state of research concerning two key challenges in biaxial tensile testing: inverse parameter identification and specimen design is reviewed.

Thom (1998) reviewed models to predict the biaxial strength of CoFRP, pointing out the lack of experimental methods available at that time to validate these models. Additionally, some possibilities for testing CoFRP under biaxial stress states, including biaxial tensile testing, were presented. Antoniou et al. (2010) modeled failure stress states in epoxy-based CoFRTS and validated their results with biaxial tensile tests. Périé et al. (2009) identified model parameters of an anisotropic damage law of carbon/carbon composites with biaxial tensile tests. Markis et al. (2007)

[1] This section contains parts of the papers "Biaxial tensile tests and microstructure-based inverse parameter identification of inhomogeneous SMC" (Schemmann et al., 2018a), "Anisotropic mean-field modeling of debonding and matrix damage in SMC composites" (Schemmann et al., 2018b), and "Cruciform specimen design for biaxial tensile testing of SMC composites" (Schemmann et al., 2018c).

provided an overview of biaxial tensile testing devices and methods that can be applied to FRP.

For the heterogeneous stress and strain state and multiaxial loading even in cruciform specimen with homogeneous properties, it is, in general, not possible to analytically identify the material parameters of a constitutive equation. Therefore, an inverse parameter identification method is needed. Schnur and Zabaras (1992) introduced a coupled procedure that links finite element simulations to an optimization method for inverse parameter identification. Through the minimization of a goal function that describes the deviation of a simulation and the experiment, the unknown material parameters are identified.

In the past, many approaches were suggested for the optimization of the goal function. A key benefit of gradient free procedures such as neural network computation (Chamekh et al., 2009) and evolutionary algorithms (Pan et al., 2010) is that no additional model information is needed. These procedures, however, usually result in comparably high computational cost. An overview of different gradient methods for inverse parameter identification is given by Ponthot and Kleinermann (2006). The required gradient can either be estimated by a finite difference scheme (Schnur and Zabaras, 1992; Cooreman et al., 2007) or calculated analytically (Mahnken and Stein, 1996a;b). The latter one requires modifications in the finite element code. The development of full-field measurement techniques allows for a parameter identification with inhomogeneous displacement and strain fields. Lecompte et al. (2007) identified the anisotropic elastic behavior of glass fiber-reinforced epoxy. Through using a goal function that compares strain fields, the needed gradient for the optimization method can be obtained analytically using the constitutive equations without making adjustments to the finite element code.

In the past, many attempts have been made to find an appropriate cruciform specimen design, demonstrating that the specimen design is one of the most challenging aspects of the biaxial testing. Several authors have proposed specimen designs for specific applications and materials, varying the cut shape, the type of the thickness-reduced area and the type of slits in the specimen arms which serve to reduce undesired lateral constraints on the strain in the center region of the specimen (Ohtake et al., 1999). Deng et al. (2015), Kuwabara et al. (1998), and Makinde et al. (1992) optimized specimens for metals. Andrusca et al. (2014) evaluated different specimens with a circular thickness-reduced area in the center, that contained smooth, spline-type thickness transition. Escárpita et al. (2009) and Gutiérrez et al. (2016) evaluated an enhanced specimen design with a thickness-reduced rectangular center area, whereas the diagonals of the rectangle are aligned with the cruciform specimen's axes. Demmerle and Boehler (1993) compared different proposed specimen designs and chose the specimen of Kelly (1976) with slits and a thickness reduced area to perform an optimization for isotropic materials. Boehler et al. (1994) investigated anisotropic sheet metals with this specimen type. Hoferlin et al. (2000) presented an alternative specimen design with slits and reinforced arms. Hannon and Tiernan (2008) reviewed planar biaxial tensile test systems for sheet metals. ISO 16842 (2014) is the first standardization in this field but applies to sheet metals only. Green et al. (2004) proposed a sandwich design for aluminum sheet alloys in which the sample sheet is bonded by an adhesive between two face sheets, while leaving the center region free on both sides.

Other authors conducted research to find a suitable specimen for composites and polymers. Smits et al. (2006) and Van Hemelrijck et al. (2007) investigated the influence of parameters like the radius of the corner fillet and the thickness and the geometry of the biaxially loaded area by finite element simulations and experiments. This investigation led to a suitable specimen for fiber-reinforced composite laminates with a reduced thickness area in the center region of the specimen, in combination with a fillet corner between arms. Makris et al. (2010) and Makinde et al. (1992)

optimized the specimen shape by a numerical optimization technique with a parametric finite element model. Lamkanfi et al. (2010) and Gower and Shaw (2010) showed that geometrical discontinuities like the transition zone to the tapered thickness area have a major influence on the strain distribution leading to premature failure. Serna Moreno et al. (2013) compared specimen with arms of different widths for chopped glass-reinforced polyester. They presented a specimen that was suitable to achieve failure in the center area in different loading cases but pointed out that there are still problems such as stress concentrations outside the center area. This glance in the literature shows that despite the promising improvements already achieved, there are still many challenges ahead in the search for suitable cruciform specimen designs for composites.

1.2.2 Homogenization of discontinuous fiber-reinforced polymers

In this section, the current state of research in micromechanical homogenization of heterogeneous materials is discussed. Special attention is given to DiCoFRP, whereas the majority of previous work has focused on DiCoFRTP. Because the goals are two-scale finite element simulations, the focus lies on the computationally efficient mean-field models. Many mean-field homogenization schemes are based on the Eshelby (1957) solution and can be classified in bounding and estimating methods (Müller, 2016). The fundamentals of such approaches are extensively addressed in textbooks by Mura (1987), Torquato (2002), and Nemat-Nasser and Hori (1999).

Bounding homogenization methods describe the upper or lower bounds of the effective material properties of a given microstructure. The well-known Voigt and Reuss bounds, originally introduced by Voigt (1889) and Reuss (1929), define the upper and lower bounds of effective material properties solely on the basis of the constituent volume fraction.

Voigt and Reuss bounds are therefore bounds of first order. The Hashin-Shtrikman bounds, in contrast, are bounds of second order that are based on a variational formulation (Hashin and Shtrikman, 1962; Willis, 1977; Walpole, 1981). Willis (1977; 1981) applied the Hashin-Shtrikman bounds to composites consisting of a matrix with ellipsoidal inclusions. Böhlke and Lobos (2014) applied the Hashin-Shtrikman bounds to anisotropic materials. In contrast to a homogeneous comparison material in the traditional Hashin-Shtrikman approach, the consideration of a heterogeneous comparison material is called linear comparison composite approach (Ponte Castañeda and Suquet, 1998; Ponte Castañeda and Willis, 1999). This comparison composite adds an additional degree of freedom, which can be exploited advantageously. Furthermore, Kröner (1977) derived third order bounds, which are narrower than the second order Hashin-Shtrikman bounds.

Estimating homogenization methods include estimation procedures for the effective material behavior. The assumption of a "dilute distribution" leads to one of the simplest homogenization schemes. The localization relation only considers inclusions in an infinite matrix, which neglects interaction between the inclusions. Thus, the estimation quality significantly drops with increasing inclusion volume fractions. For high inclusion volume fractions, the assumption of dilute distributions may violate Voigt or Reuss bounds, thus leading to unphysical results. Maxwell (1873) and Einstein (1905) originally introduced the dilute distriubtion assumption to estimate the effective conductivity and viscosity of spheres dispersed in a viscous fluid. Benveniste et al. (1991) and Tucker III and Liang (1999) approximated the stiffness of composites with dilute distribution estimates.

The Mori-Tanaka homogenization estimate is introduced in more detail in Section 3.3.2. Mori and Tanaka (1973) assumed that the inclusion strain is estimated by the strain localization of the single inclusion problem with the phase-averaged matrix strain instead of the macroscopic strain and followed the procedure to derive an expression for effective material behavior. The advantages of the Mori-Tanaka homogenization scheme include its simple derivation and implementation, its acceptance in academia and industry, and the large number of published validations. However, the violations of the main symmetry condition of the effective stiffness for anisotropic phases and the lack of a derivation with a variational formulation are significant drawbacks.

The basis of the self-consistent scheme is the dilute distribution scheme, with the modification that the inclusion is embedded in an infinite matrix consisting of the effective material (Hill, 1965a; Budiansky, 1965; Kröner, 1977). With linear elasticity, the self-consistent scheme defines a nonlinear implicit equation that defines the effective material behavior. A linearized formulation of the self-consistent scheme allows one to consider nonlinear material behavior (Hill, 1965b; Suquet, 1995).

A major drawback of the self-consistent scheme is the missing interaction between the inclusion and the matrix material, since the inclusion is directly embedded in the effective material. The generalized self-consistent scheme is based on a three-phase model. Here, the inclusion is embedded in a matrix cell, which is in turn embedded in the effective material. Christensen and Lo (1979) and Hashin (1968) applied the generalized self-consistent scheme to composite materials and, therefore, drew significant attention to the it. Benveniste (2008) reviewed the generalized self-consistent scheme in great detail and presented an alternative formulation.

The interaction direct derivative method, presented by Zheng and Du (2001) and Du and Zheng (2002), is based on an approximation of the generalized self-consistent estimate. One advantage of the interaction direct derivative method lies in its explicit structure, it is valid for multi-phase composites with different material symmetries and distributions. Müller et al. (2015a) present an application of the interaction direct derivative and self-consistent estimate to short-fiber-reinforced polypropylene.

Schemmann et al. (2015b) present a validation of interaction direct derivative estimate with long-fiber thermoplastics in biaxial tensile tests. For the special case in which the matrix cell and the inclusion have similar geometries and isotropic constituents, the interaction direct derivative and Mori-Tanaka schemes evaluate the same effective material behavior (Du and Zheng, 2002).

Pierard et al. (2004) and Doghri and Friebel (2005) homogenized composites in a two-step scheme. In the first step, the microstructure domain is divided into subdomains. For discontinuous fiber-reinforced composites, Ogierman and Kokot (2017) generated subdomains by discretizing fiber directions, thus obtaining subdomains with unidirectional fiber orientations. The subdomains or pseudo grains are then homogenized separately. In the second step, these subdomains are homogenized to yield an effective material behavior. In the literature the two-step scheme is also known as the pseudo-grain method.

In contrast to mean-field homogenization schemes, the basis of full-field homogenization schemes is a spatially-resolved heterogeneous microstructure. A representative volume element (RVE) is discretized and subjected to external loads. The effective material behavior is obtained by the RVE response of resulting boundary value problems. Typical numerical full-field homogenization schemes are based on the finite element method (FEM), fast Fourier transformation (FFT) approaches, or the phase-field method (PFM).

Suquet (1987) and Moulinec and Suquet (1994) applied an FFT approach to solve Lipmann-Schwinger-type equations. A main advantage of full-field FFT over traditional FEM is the reduced memory demand. The voxel-based formulation allows for direct application to segmented microstructure images, such as μCT datasets. Müller et al. (2015b) compared full-field FFT solutions with mean-field schemes (self-consistent and interaction direct derivative estimate).

1.2.3 Damage modeling

This section provides an overview of damage modeling and follows with a detailed presentation of the current state of research in damage characterization and mean-field damage modeling of CiDoFRTS.

Before the modeling of stiffness degradation, damage was seen as a logical variable in terms of failed or sound material, as investigated, for example, by Weibull (1951). Kachanov (1958) and Rabotnov (1968) modeled progressive creep damage and developed the effective stress concept. The two textbooks Lemaitre (1996) and Krajcinovic (1984) served as the basis for much research in continuum damage mechanics. The challenge of softening materials in finite element analysis was reviewed in detail by Belytschko et al. (1986). Forest and Lorentz (2004) reviewed localization issues and regularization methods.

Meraghni and Benzeggagh (1995) investigated damage propagation in randomly-oriented, discontinuous, fiber-reinforced composites. Their experimental studies involving the amplitude analysis of acoustic emission signals and microscopic observations revealed two dominant damage mechanisms: matrix damage and interface damage. Several other authors (e.g., Jendli et al. (2004); Ben Cheikh Larbi et al. (2006); Fitoussi et al. (2013)) have confirmed that fiber-matrix interface debonding is the primary and matrix cracking the secondary damage mechanism in SMC.

Experimental characterization of the interface strength is a challenging field. Many authors (e.g., Favre and Jacques (1990); Desarmot and Favre (1991); Tandon et al. (2002); Koyanagi et al. (2012)) have investigated interface strength using pull-out, push-out, and micro-tensile techniques or via fragmentation tests. Hour and Sehitoglu (1993) and Dano (2006), e.g., performed experimental studies with focus on the development of damage in glass fiber-reinforced composites and Fitoussi et al. (2005) conducted experiments specifically on the mechanical behavior of SMC.

Jendli et al. (2004) qualitatively analyzed the influence of the strain rate on damage threshold and accumulation. Performing monotonic and interrupted tensile tests at different strain rates, Jendli et al. (2005) showed that both damage onset and kinetics are sensitive to the strain rate, such that the interface failure strength increases with increasing strain rate. Similar findings were obtained by Fitoussi et al. (2013) and Shirinbayan et al. (2015).

Along with their experimental findings, Fitoussi et al. (1996a) also proposed a micromechanical model based on an equivalent, anisotropic inhomogeneity approach for damaged fibers. Their fiber-matrix interface debonding model is based on a criterion with linear coupling of the local shear and normal stress on the interface (Fitoussi et al., 1996b). This work was followed by an extension that considered local fluctuations of strain and stress, and a probabilistic interface-strength distribution (Fitoussi et al., 1998). Meraghni et al. (1996) developed a similar model that combined a microcrack density parameter with fiber-matrix decohesion in order to decrease the fiber strain localization tensor. Derrien et al. (2000) further developed the probabilistic interface-damage model introduced by Fitoussi et al. (1998). Here, the approach of replacing fibers having damaged interfaces with an equivalent, anisotropic, undamaged inhomogeneity or matrix material were validated experimentally. Desrumaux et al. (2000) extended the statistical representation of failure to each constituent (fibers, matrix, and interface). Here, the interface damage was implemented in such a way that the interface of each fiber is experiencing damage represented by the introduction of directionally-dependent matrix cracks. Later, Desrumaux et al. (2001) introduced a two-step homogenization damage model for a randomly-oriented fiber composite based on a numerically determined Eshelby tensor. A comparable two-step homogenization framework was pursued by Jendli et al. (2009) and Kammoun et al. (2015), who followed approaches for interfacial decohesion and pseudo-grain sub-regions. Meraghni et al. (2002) further developed the probabilistic strength model. Analogously, Guo et al. (1997) introduced a damageable-elastic law for randomly-reinforced composites based on a two-scale approach. Nguyen and Khaleel (2004) developed a matrix degradation model based on the experimental findings of Meraghni and Benzeggagh (1995). Here, the macroscopic stiffness of the randomly-oriented composite is calculated by an orientation average over aligned fibers, and a thermodynamically consistent damage evolution law predicts the stiffness reduction. Baptiste (2003) proposed a model that captures the inelastic behavior of a composite due to plasticity, viscosity, and damage. The damage to matrix, reinforcement, and interface is considered utilizing the probabilistic approach for interface-strength developed by Fitoussi et al. (1998). Lee and Simunovic (2001) evolved a model to predict the elasto-plastic-damage behavior of a ductile matrix composite containing aligned fibers. They combined an associative flow rule and a hardening law with evolutionary interfacial debonding. In their model, partially debonded fibers are replaced by equivalent, perfectly-bonded fibers. Additionally, they extended their model to treat random fiber orientations (Lee and Simunovic, 2000). A similar model was developed by Ju and Lee (2000) for a three-phase composite. Here, completely debonded fibers are treated as voids within the three-phase homogenization scheme. Using an effective yield criterion, an associative plastic flow rule, and a hardening law, the macroscopic mechanical behavior predicted by the model fits the experimental observations. Zaïri et al. (2008) described interface debonding as nucleation and the growth of voids. They combined a critical void volume criterion with a vanishing element technique to capture damage accumulation and failure. Ben Cheikh Larbi et al. (2006) investigated the elastic behavior of SMC under cyclic loading. They found appropriate parameters for the two-scale probabilistic damage model by evaluating fatigue tests via scanning electron microscopy. Yang et al. (2012) introduced a phenomenological damage model based on two coupled damage variables to capture matrix cracking and interface debonding. A von Mises type criterion and a cohesive zone model are applied to capture the two damage mechanisms. Notta-Cuvier et al. (2014) developed a model to describe interface debonding at fiber head surfaces in injection-molded, short-fiber-reinforced composites. The model for interface debonding utilizes the accumulation of voids and phenomenological parameters. Schulenberg et al. (2017) presented a computationally efficient damage model for DiCoFRTP. Seelig (2008) suggested a modeling approach for CoFRTP with a visco-plastic matrix behavior. The cohesive surface formulation could be applied to CoDiCoFRP.

Regarding full-field simulations, Spahn et al. (2014) obtained an FFT solution for damage evolution in short-fiber-reinforced thermoplastics. The simulation of crack propagation in multi-phase materials with the phase field method is still in an early stage of research but first results (Biner and Hu, 2009; Schneider et al., 2016) are promising.

1.3 Outline of the thesis

This thesis consists of two chapters that treat the fundamentals of the subject, followed by three chapters that are based on published journal publications. Additionally, the key achievements in terms of research progress beyond the current state of the art are highlighted.

Chapter 2 introduces the SMC manufacturing process and the resulting microstructure. In particular, it focuses on the introduction of the fiber orientation distribution function and fiber orientation tensors.

Chapter 3 provides the continuum mechanical basis for the following chapters. Here, the kinematics and relevant balance equations in the context of small deformations are introduced. An introduction to micromechanical modeling of heterogeneous materials covers effective properties, localization relations, and the fundamentals of the Mori-Tanaka homogenization scheme.

Chapter 4 presents a parameter identification scheme in macroscopically inhomogeneous linear elastic SMC. The assumption of linearity between the fiber orientation and the stiffness reduces the degrees of freedom of the parameter identification to five. Previous work applied the linearity of the orientation tensor and effective stiffness in homogenization schemes. The novelty lies in the application to DiCoFRP, as well as an implementation of an inhomogeneous parameter identification.

Chapter 5 discusses different cruciform specimen designs for damage characterization in SMC under biaxial loadings. Various (material-specific) requirements for an optimal specimen design are defined. One key challenge represents the achievement of a high strain level in the center region of the cruciform specimen while preventing premature failure in the clamped specimen arms. The added value of this chapter lies in the consideration of DiCoFRP in biaxial tensile tests as well as the unique manufacturing technology and strategy by combining co-molding of CoDiCoFRTS and gentle milling.

Chapter 6 presents an anisotropic, micromechanical damage model for SMC. The damage model captures the interface damage and matrix damage in a mechanism-based mean-field model. The model is validated under two stress states and with two different matrix systems. In comparison to the extensive previous work in damage modeling of DiCoFRTS, the added value of this model lies in the consideration of inhomogeneous stress distribution on the fiber-matrix interfaces, the general formulation of equivalent interface stress, the validated materials and stress states, and the extensive visualization of the anisotropic damage evolution.

1.4 Notation

A direct tensor notation is followed throughout the text. Tensor components are expressed by latin indices, and Einstein's summation convention is applied. Components of vectors and tensors refer to the orthonormal basis $\{\boldsymbol{e}_1, \boldsymbol{e}_2, \boldsymbol{e}_3\}$. Vectors and second-order tensors are denoted by lower case and upper case bold letters, respectively (e.g., $\boldsymbol{a}$ and $\boldsymbol{A}$). Fourth-order tensors are denoted by, e.g., $\mathbb{A}, \mathbb{B}$. Additionally, higher-order tensors are written as $\mathbb{A}_{\langle\alpha\rangle}$, where α indicates the tensor rank. The composition of two second-order or two fourth-order tensors is formulated by $\boldsymbol{AB}$ and $\mathbb{AB}$. A linear mapping of vectors by second-order tensors and second-order by a fourth-order tensor is written as $\boldsymbol{a} = \boldsymbol{Cb}$ and $\boldsymbol{A} = \mathbb{C}[\boldsymbol{B}]$. The scalar product is denoted by $\boldsymbol{A} \cdot \boldsymbol{B}$. The dyadic product operator $\otimes$ is defined as $(\boldsymbol{A} \otimes \boldsymbol{B})[\boldsymbol{C}] = (\boldsymbol{B} \cdot \boldsymbol{C})\,\boldsymbol{A}$, higher-order dyadic products of the same tensor are indicated by $\boldsymbol{n}^{\otimes\alpha} = \boldsymbol{n} \otimes \cdots \otimes \boldsymbol{n}$, where $\boldsymbol{n}^{\otimes\alpha}$ is a tensor with the rank α times the rank of $\boldsymbol{n}$. The Rayleigh product $\boldsymbol{Q} \star \mathbb{A} = A_{ijkl}(\boldsymbol{Q}\boldsymbol{e}_i) \otimes (\boldsymbol{Q}\boldsymbol{e}_j) \otimes (\boldsymbol{Q}\boldsymbol{e}_k) \otimes (\boldsymbol{Q}\boldsymbol{e}_l)$ describes, e.g., the rotation of a fourth order tensor with the orthogonal rotation tensor $\boldsymbol{Q}$. The identity on symmetric fourth-order tensors is denoted by $\mathbb{I}^{\mathrm{s}}$. Completely symmetric and traceless, i.e. irreducible tensors are denoted with a prime, e.g., $\boldsymbol{A}'$. Column vectors and matrices are identified via underscores: e.g., $\underline{p}$, $\underline{\underline{A}}$.

Chapter 2

Microstructure of SMC

2.1 Manufacturing process

Since the SMC manufacturing process significantly influences the microstructure and mechanical behavior of the final SMC parts, an overview of this manufacturing process is given first. The focus of this section lies on pre-impregnated fiber (prepreg) production in an SMC line and the subsequent compression molding. Preliminary steps, such as thermoset or glass fiber production and post-processing, such as milling or water jet cutting, are not covered here. Nor is the addition of filler particles addressed. For more details concerning SMC manufacturing and composite manufacturing in general, the reader is referred to the works of Henning and Moeller (2011), Advani and Hsiao (2012), and Cherif (2016).

Figure 2.1 shows an SMC production line, in which the main production direction is from right to left. One doctor box spreads resin, such as epoxy or polyurethane, on the lower film. The chopper cuts the continuous fiber rovings into 2.54cm long pieces. Typical SMC fiber materials include carbon and glass. The chopped fibers fall onto the lower resin layer. The initial orientation distribution of the chopped fibers is typically assumed to be planar isotropic. A second film with resin is guided on top of the lower resin and fibers. This material, now consisting of two resin and one fiber layer covered with films on the top and bottom, is fed into the impregnation belt. The curved path applies relative motion between the constituents, which improves impregnation. A roller winds up the prepregs for better handling and storage. The storage of prepregs at a specific temperature partly cures the thermoset matrix, a process also known as pre-curing.

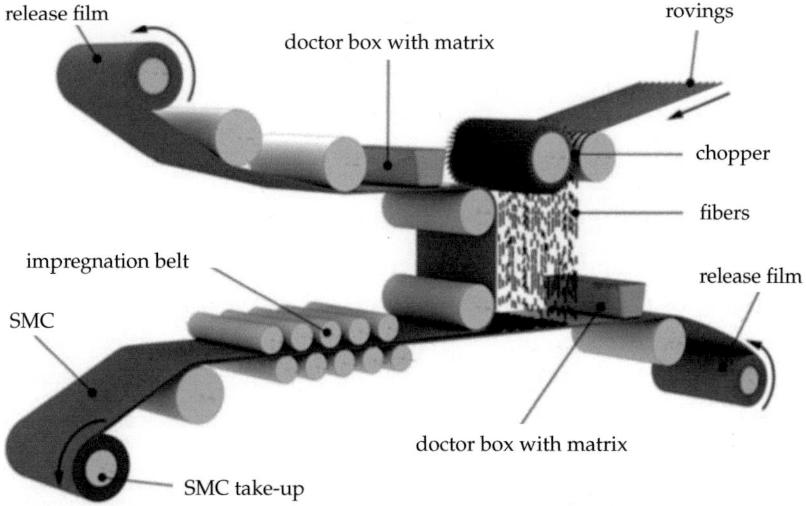

Figure 2.1: SMC prepreg manufacturing line (original image from Cherif (2016) modified by Bücheler (2017))

After the pre-curing period, the SMC prepregs are cut and stacked so as to yield the necessary size and thickness of the initial charge. Figure 2.2 shows the placement of this initial charge within the tempered (typically $120 - 160°C$) mold. The closing of the press fills the mold. In contrast to thermoplastic compression molding, thermoset composites do not exhibit fountain flow. The flow in the mold is more plug-like; the center of the material experiences little shear and deforms primarily in biaxial extension, since the outer (hot) layers have a significantly lower viscosity (Davis et al., 2003). The mold-flow induces an evolution in the fiber orientation distribution, generally leading to a process-dependent inhomogeneous fiber orientation distribution. Jeffery (1922) laid the foundation for the modern modfilling analysis, as he described the motion of a rigid ellipsoid in shear flow. Other authors (e.g. Advani and Tucker III (1987); Zheng et al. (2011); Bertóti and Böhlke (2017)) mathematically modeled the mold-filling process, including the evolution of the fiber orientation distribution.

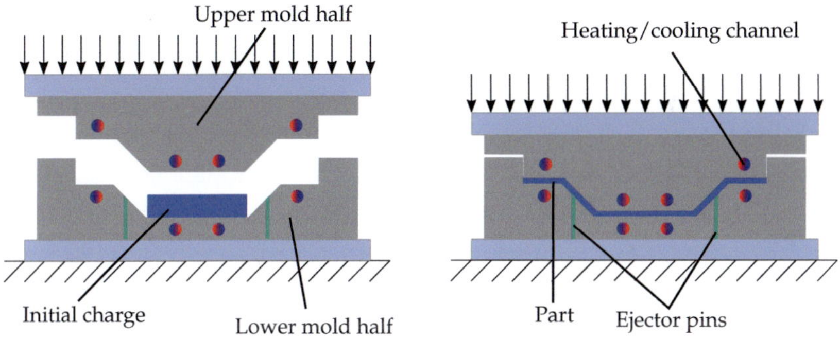

Figure 2.2: Compression molding of an SMC part

After the mold is completely filled, the thermoset matrix cures in the closed mold, typically for two to four minutes. This assures low cycle times and makes SMC suitable for mass production. When the mold is opened, ejector pins remove the final SMC part from the mold.

2.2 Classification and general model assumptions

A heterogeneous material is a material that is composed of domains of different materials (phases). Heterogeneous materials include wood, human tissue, reinforced concrete, polycrystalline metals, and fiber-reinforced polymers.

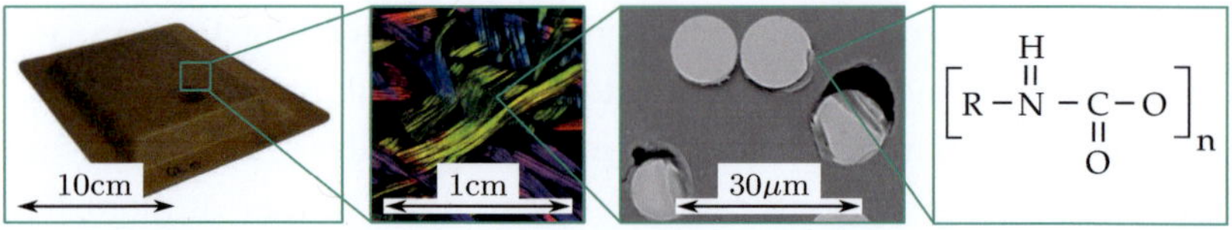

Figure 2.3: Scales of an SMC component, fiber roving and filament microstructure, and molecular structure of polyurethane resin (GRK 2078 demonstrator manufactured by David Bücheler (Fraunhofer ICT Pfinztal, Germany), μCT scan of SMC microstructure captured and analyzed by Pascal Pinter (IAM-WK, KIT), REM image of fiber filaments in matrix provided by Michael Schober (Fraunhofer IWM Freiburg, Germany))

Figure 2.3 illustrates SMC at different scales. Typical components (left image) are shell-like structures with in-plane dimensions typically ranging from a few centimeters to two meters. Thicknesses typically range from two to four millimeters. There is no scale separation to the adjacent image, which is a colored, micro-computed tomography (μCT) scan of fiber bundles. The fibers are approximately $25.4\mathrm{mm}$ long. The next, grey scale image is a scanning electron microscope capture that shows individual filaments of a mechanically-damaged fiber-reinforced polymer. Most filament diameters are in the range of several micrometers. The molecular scale provides insight into much smaller scales, as defined by the polymer macromolecule and molecular structure of the fibers and their sizing.

Drugan and Willis (1996) define a representative volume element (RVE) by its constitutive behavior with respect to the macroscopic material behavior. Another definition (Drugan and Willis, 1996) states that the smallest RVE is the one that is sufficiently large to be statistically representative of the composite, and therefore efficiently includes a sampling of all possible microstructural configurations of the composite. A main part of this thesis focuses on the mean-field homogenization from the microscale (middle two images of Figure 2.3) to the component scale or macroscale (left image in Figure 2.3). Such traditional homogenization schemes are based on a separation of scales between a microstructure RVE and the component dimensions or macroscale. Such an RVE and such a scale separation do not exist in SMC. The application of mean-field homogenization schemes in SMC is, however, state of the art and has lead to satisfying results in many two-scale structural simulations. Here, the applicability of mean-field homogenization schemes is assumed and the missing scale separation is discussed in more detail in Section 3.3.1. The remainder of this chapter introduces the mathematical fundamentals needed to describe a fibrous microstructure.

The total volume of any heterogeneous microstructure v is given by the sum of all phase volumes v_α

$$v = \sum_{\alpha=1}^{N_\mathrm{P}} v_\alpha. \tag{2.1}$$

Here, N_P denotes the number of discrete phases. The phase volume fraction c_α represents the share of the specific phase volume relative to the total volume

$$c_\alpha = \frac{v_\alpha}{v}. \tag{2.2}$$

Torquato (2002) states that the realization of each sample is the result of a stochastic process determining the microstructure. Therefore, the mechanical quantity ξ also depends on the realization $\omega \in \Omega$, with Ω representing the set of all possible realizations, also known as the ensemble. The influence of a statistical process can be eliminated by the ensemble average $\langle \cdot \rangle_\Omega$

$$\langle \xi\left(\boldsymbol{x}, t\right) \rangle_\Omega = \lim_{N \to \infty} \frac{1}{N} \sum_{i=1}^{N} \xi\left(\boldsymbol{x}, t, \omega_i\right). \tag{2.3}$$

The ergodic hypothesis implies that the ensemble average is equivalent to the volume average $\langle \cdot \rangle_{\omega_0}$ in one realization ω_0 in the infinite-volume limit

$$\langle \xi\left(\boldsymbol{x}_0, \omega\right) \rangle_\Omega = \langle \xi\left(\boldsymbol{x}, \omega_0\right) \rangle_{\omega_0} \quad \omega_0 \to \infty. \tag{2.4}$$

For a more detailed description of the microstructure, the indicator function $\mathcal{I}_\alpha(\boldsymbol{x}, \omega)$ is introduced. The indicator function equals one if the position vector $\boldsymbol{x}$ points in phase α, and zero otherwise

$$\mathcal{I}_\alpha(\boldsymbol{x}, \omega) = \begin{cases} 1 & \forall \boldsymbol{x} \in v_\alpha(\omega) \\ 0 & \text{otherwise} \end{cases}. \tag{2.5}$$

The indicator functions are complete, idempotent and bi-orthogonal

$$\sum_{\alpha=1}^{N} \mathcal{I}_\alpha(\boldsymbol{x}, \omega) = 1 \qquad \text{(completeness)} \tag{2.6}$$

$$(\mathcal{I}_\alpha(\boldsymbol{x}, \omega))^2 = \mathcal{I}_\alpha(\boldsymbol{x}, \omega) \qquad \text{(idempotence)} \tag{2.7}$$

$$\mathcal{I}_\alpha(\boldsymbol{x}, \omega)\, \mathcal{I}_\beta(\boldsymbol{x}, \omega) = 0 \quad \forall \alpha \neq \beta \qquad \text{(bi-orthogonality).} \tag{2.8}$$

The n-point correlation function describes the probability that phase α is found at points $\boldsymbol{x}_1, \cdots, \boldsymbol{x}_n$

$$S_n^\alpha(\boldsymbol{x}_1, \ldots, \boldsymbol{x}_n) = \langle \prod_{k=1}^{n} \mathcal{I}_\alpha(\boldsymbol{x}_k, \omega) \rangle. \tag{2.9}$$

The one-point correlation function $S_1^\alpha(\boldsymbol{x}_1) = \langle \mathcal{I}_\alpha(\boldsymbol{x}_1, \omega) \rangle_\Omega$ describes the probability of finding phase α at point $\boldsymbol{x}$ and can be interpreted as a position-dependent fiber volume fraction (Jöchen, 2013).

A material with a translation-invariant n-point correlation function is called statistically homogeneous

$$S_n^\alpha(\boldsymbol{x}_1, \ldots, \boldsymbol{x}_n) = S_n^\alpha(\boldsymbol{x}_1 + \Delta \boldsymbol{x}, \ldots, \boldsymbol{x}_n + \Delta \boldsymbol{x}). \tag{2.10}$$

Here, Δx denotes an arbitrary vector that describes the translation. A statistically isotropic material is characterized by a rotation invariant n-point correlation function

$$S_n^{\alpha}\left(\boldsymbol{x}_1, \ldots, \boldsymbol{x}_n\right) = S_n^{\alpha}\left(\boldsymbol{Q}\boldsymbol{x}_1, \ldots, \boldsymbol{Q}\boldsymbol{x}_n\right) \quad \boldsymbol{Q} \in Orth. \tag{2.11}$$

Here, $\boldsymbol{Q}$ denotes an arbitrary orthogonal rotation tensor.

Hereafter SMC is modeled as a two-phase microstructure that consists of an isotropic thermoset matrix M and cylindrical reinforcing long glass or carbon fibers F, whereas there is no scale separation due to the ratio of fiber length to component thickness. Due to the process-induced fiber orientation distribution, SMC is neither statistically homogeneous, statistically isotropic, nor ergodic.

2.3 Fiber orientation distribution function

This section introduces the fiber orientation distribution function, which describes the fiber orientation statistics in a one-point correlation function. Hereafter, the fibers are assumed to be straight. For investigations into the effect of fiber curvature, the reader is referred to Kosker and Akbarov (2003) and Bapanapalli and Nguyen (2008). First, the unit vector $\boldsymbol{n}$ is defined as the orientation of a single fiber (Figure 6.3). The fiber orientations $\boldsymbol{n}$ and $-\boldsymbol{n}$ are equivalent.

Figure 2.5 shows a micro-computed tomography scan of a typical SMC microstructure. The wide range of colors in this image, which represent different fiber orientations $\boldsymbol{n}$, highlights the need to describe the fiber orientation statistically. Here, $\varphi = \angle\left(\boldsymbol{e}_1, \left(\boldsymbol{n} \cdot \boldsymbol{e}_1\right) \boldsymbol{e}_1 + \left(\boldsymbol{n} \cdot \boldsymbol{e}_2\right) \boldsymbol{e}_2\right)$ describes the angle between the $\boldsymbol{e}_1$ direction and the fiber projection on the $\boldsymbol{e}_1$-$\boldsymbol{e}_2$-plane. The fiber orientation analysis was performed following the procedure of Pinter et al. (2018).

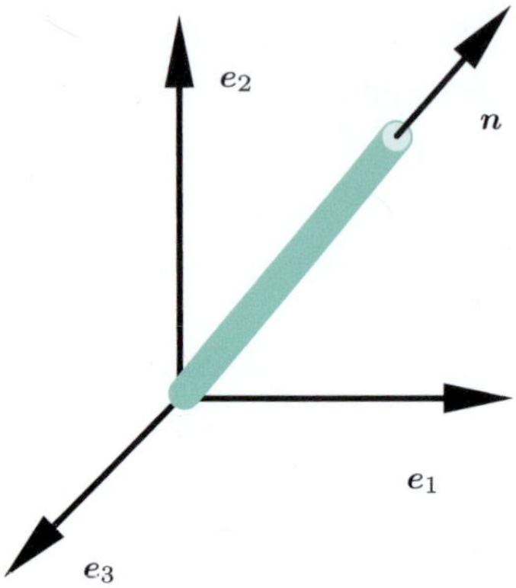

Figure 2.4: A single straight fiber with orientation $\boldsymbol{n}$

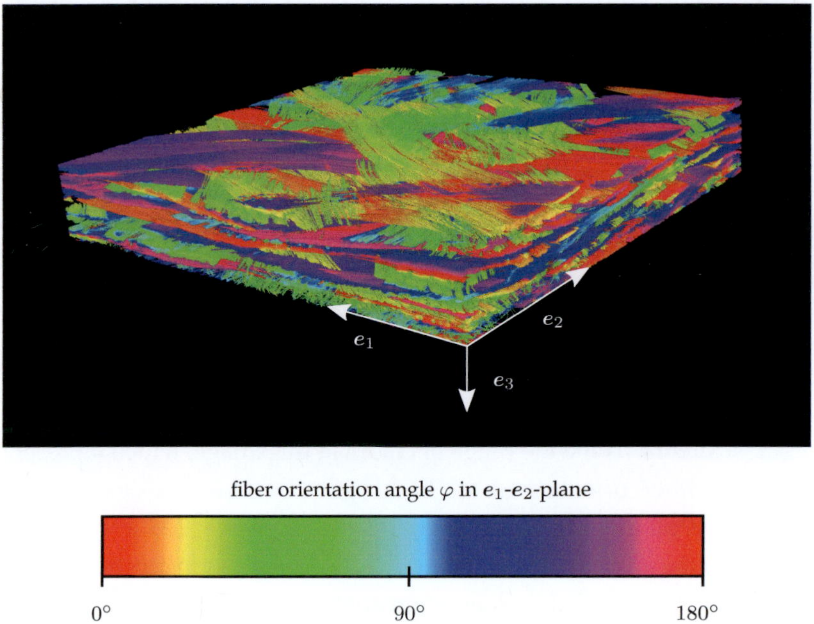

Figure 2.5: A micro-computed tomography (μCT) scan of a SMC microstructure (UPPH resin, fiber volume fraction $c_{\mathrm{F}} = 23\%$). The specimen dimensions are $10\mathrm{mm} \times 10\mathrm{mm} \times 3\mathrm{mm}$ (Pinter, 2017).

The fiber orientation and length distribution function specifies the volume fraction $\mathrm{d}v/v$ of fibers with orientation $\boldsymbol{n}$ and length l with respect to all fibers

$$\frac{\mathrm{d}v}{v}(\boldsymbol{n}, l) = f(\boldsymbol{n}, l)\,\mathrm{d}S. \tag{2.12}$$

Here, $\mathrm{d}S$ is the surface element on the unit sphere which is defined as $S := \{\boldsymbol{n} \in \mathbb{R}^3 : \|\boldsymbol{n}\| = 1\}$. A parametrization of $\mathrm{d}S$ in spherical coordinates with the azimuth angle ϑ and the polar angle ϕ can yield (Böhlke et al., 2010)

$$\mathrm{d}S = \frac{\sin\vartheta}{4\pi}\mathrm{d}\vartheta\mathrm{d}\phi. \tag{2.13}$$

The corresponding integral of a quantity $\psi(\boldsymbol{n}) = \hat{\psi}(\vartheta, \phi)$ on the unit sphere surface is given by

$$\int_S \psi(\boldsymbol{n})\,\mathrm{d}S = \int_{\phi=0}^{2\pi}\int_{\vartheta=0}^{\pi} \hat{\psi}(\vartheta, \phi)\,\frac{\sin\vartheta}{4\pi}\mathrm{d}\vartheta\mathrm{d}\phi. \tag{2.14}$$

The fiber orientation and length distribution are generally dependent on each other. In case of SMC, with typically almost uniform length distribution in the prepreg, Motaghi and Hrymak (2017) did not find evidence for fiber break during the manufacturing process. DiCoFRTP is, however, known to experience fiber break in the manufacturing process and thus the fibers with different aspect ratios experience different re-orientating forces. Integrating over all fiber lengths yields the fiber orientation distribution function (FODF). The FODF specifies the volume fraction $\mathrm{d}v/v$ of fibers with orientation $\boldsymbol{n}$ with respect to all fibers (see, e.g. Zheng and Du (2001); Müller and Böhlke (2016))

$$\frac{\mathrm{d}v}{v}(\boldsymbol{n}) = f(\boldsymbol{n})\,\mathrm{d}S. \tag{2.15}$$

The FODF is symmetric, normalized, and non-negative:

$$f(\boldsymbol{n}) = f(-\boldsymbol{n}) \qquad \text{(symmetric)} \tag{2.16}$$
$$\int_S f(\boldsymbol{n})\mathrm{d}S = 1 \qquad \text{(normalized)} \tag{2.17}$$
$$f(\boldsymbol{n}) \geq 0 \qquad \text{(non-negative).} \tag{2.18}$$

The fiber distribution function (FDF) $h\left(\boldsymbol{n}\right)$ specifies the fraction of (intact) fibers $\mathrm{d}N$ with orientation $\boldsymbol{n}$ with respect to all fibers:

$$\mathrm{d}N(\boldsymbol{n}) = h\left(\boldsymbol{n}\right)\mathrm{d}S. \tag{2.19}$$

The FDF is symmetric and non-negative. Integrating over the unit sphere S yields the total number of fibers N:

$$h(\boldsymbol{n}) = h(-\boldsymbol{n}), \quad h(\boldsymbol{n}) \geq 0, \quad \int_S h(\boldsymbol{n})\mathrm{d}S = N. \tag{2.20}$$

The relation between the FODF and the FDF is given by

$$f(\boldsymbol{n}) = \frac{h(\boldsymbol{n})}{N}. \tag{2.21}$$

2.4 Fiber orientation tensors

2.4.1 Tensors of the first kind

Advani and Tucker III (1987) introduced the fiber orientation tensors of second order $\boldsymbol{N}$ and fourth order $\mathbb{N}$ as follows:

$$\boldsymbol{N} = \int_S f(\boldsymbol{n})\boldsymbol{n}\otimes\boldsymbol{n}\mathrm{d}S, \quad \mathbb{N} = \int_S f(\boldsymbol{n})\boldsymbol{n}\otimes\boldsymbol{n}\otimes\boldsymbol{n}\otimes\boldsymbol{n}\mathrm{d}S. \tag{2.22}$$

Higher order fiber orientation tensors of nth-order $\mathbb{N}_{\langle n \rangle}$ (where n is even and larger than four) are defined analogously

$$\mathbb{N}_{\langle n \rangle} = \int_S f(\boldsymbol{n}) \boldsymbol{n}^{\otimes n} \mathrm{d}S. \tag{2.23}$$

Kanatani (1984) calls these orientation tensors of the first kind. Equation (2.16) implies that $\boldsymbol{N}$ is completely symmetric and normalized. Higher order tensors of the same kind provide complete information about lower orders. This yields the following relation for second and fourth order orientation tensors of the first kind

$$\boldsymbol{N} = \boldsymbol{N}^{\mathsf{T}}, \quad \mathbb{N} = \mathrm{symF}\left(\mathbb{N}\right), \quad \mathrm{tr}(\boldsymbol{N}) = 1, \quad \mathbb{N}\left[\boldsymbol{I}\right] = \boldsymbol{N}. \tag{2.24}$$

Here, the operator $\mathrm{symF}\left(\cdot\right)$ describes a full symmetrization, that is, invariance with respect to all possible permutations of all indices (Ehrentraut and Muschik, 1998).

2.4.2 Tensors of the second kind

Kanatani (1984) introduced the fabric tensors of the second kind $\boldsymbol{F}$, $\mathbb{F}$, and $\mathbb{F}_{\langle n>4 \rangle}$. These are unique tensors that minimize the measure of approximation E between the approximated and exact FODF. E is defined as

$$E = \int_S \left(\mathbb{F}_{\langle n \rangle} \cdot \boldsymbol{n}^{\otimes n} - f(\boldsymbol{n})\right)^2 \mathrm{d}S \rightarrow \min. \tag{2.25}$$

For a detailed derivation of the orientation tensors of the second kind, the reader is referred to the work of Kanatani (1984) and Hotz and Schultz (2015). The second and fourth order tensors that solve the minimization can be expressed as functions of the orientation tensors of the first kind as (see Kanatani (1984), p.152)

$$\boldsymbol{F} = \frac{15}{2}\left(\boldsymbol{N} - \frac{1}{5}\boldsymbol{I}\right) \tag{2.26}$$

and

$$\mathbb{F} = \frac{315}{8}\left(\mathbb{N} - \frac{2}{3}\mathrm{symF}\left(\boldsymbol{I}\otimes\boldsymbol{N}\right) + \frac{1}{21}\mathrm{symF}\left(\boldsymbol{I}\otimes\boldsymbol{I}\right)\right). \tag{2.27}$$

Orientation tensors of the second kind are fully symmetric (Ehrentraut and Muschik, 1998) and exhibit the following properties for the traces

$$\mathrm{tr}(\boldsymbol{F}) = 3, \quad \boldsymbol{F} = \boldsymbol{F}^{\mathsf{T}}, \quad \mathrm{tr}(\mathbb{F}\left[\boldsymbol{I}\right]) = 5, \quad \mathrm{symF}\left(\mathbb{F}\right) = \mathbb{F}. \tag{2.28}$$

It is noted here that the FODF approximation given by Equation (2.25) may yield a negative FODF, thereby violating Equation (2.16). The approximation of a unidirectional fiber orientation distribution yields a negative FODF, as does the approximation of a planar orientation distribution. To approximate planar or near planar FODFs, planar fiber orientation tensors might be helpful.

2.4.3 Tensors of the third kind[1]

Kanatani (1984) introduced fabric tensors of the third kind $\mathbb{D}_{\langle n\rangle}$ as follows. Orientation tensors of the third kind are deviatoric and fully symmetric (Ehrentraut and Muschik, 1998). Therefore, the nth-order orientation tensor of the third kind has $2n+1$ independent components. These irreducible orientation tensors are traceless and fully symmetric:

$$\mathrm{tr}(\boldsymbol{D}') = 0, \quad \boldsymbol{D}' = \boldsymbol{D}'^{\mathsf{T}}, \quad \mathbb{D}'\left[\boldsymbol{I}\right] = \boldsymbol{0}, \quad \mathrm{symF}\left(\mathbb{D}'\right) = \mathbb{D}'. \tag{2.29}$$

[1] This subsection is based on the subsection "Note on Fiber Orientation Tensors" of the paper "Biaxial tensile tests and microstructure-based inverse parameter identification of inhomogeneous SMC composites" (Schemmann et al., 2018a).

By using the orientation tensors of the first kind $\mathbb{N}_{\langle n \rangle}$, the orientation tensor of the third kind $\mathbb{D}'_{\langle n \rangle}$ can be obtained as follows (Müller, 2016; Kanatani, 1984)

$$\mathbb{D}'_{\langle n \rangle} = \frac{2n+1}{2^n} \binom{2n}{n} \mathbb{N}'_{\langle n \rangle}, \tag{2.30}$$

where $\binom{n}{k}$ is the binomial coefficient. The index notation for the deviatoric part of a nth-order symmetric tensors can be found in Kanatani (1984). An explicit formulation of the fiber orientation tensors of the third kind, for first, second, and fourth order is given by (see Kanatani (1984), p.154)

$$D = 1 \tag{2.31}$$

$$\boldsymbol{D}' = \frac{15}{2} \left(\boldsymbol{N} - \frac{1}{3} \boldsymbol{I} \right) \tag{2.32}$$

$$\mathbb{D}' = \frac{315}{8} \left(\mathbb{N} - \frac{6}{7} \text{symF} \left(\boldsymbol{I} \otimes \boldsymbol{N} \right) + \frac{3}{35} \text{symF} \left(\boldsymbol{I} \otimes \boldsymbol{I} \right) \right). \tag{2.33}$$

An approximation of the FODF by orientation tensors of the third kind is given by

$$f(\boldsymbol{n}) = D + \boldsymbol{D}' \cdot \boldsymbol{n} \otimes \boldsymbol{n} + \mathbb{D}' \cdot \boldsymbol{n} \otimes \boldsymbol{n} \otimes \boldsymbol{n} \otimes \boldsymbol{n} + \dots. \tag{2.34}$$

This formulation is equivalent to the approximation with fiber orientation tensors of the second kind in Equation (2.25).

Chapter 3

Continuum mechanical foundations

3.1 Kinematics

This chapter provides an overview of the continuum mechanical foundations. First, the basics of kinematics are introduced. Then, the required balance equations are discussed. Finally, an introduction to the micromechanical modeling of composites, including the Mori-Tanaka homogenization scheme, is provided. For further information and detailed derivations with regard to kinematics and balance equations, the reader is referred the textbooks of Truesdell and Toupin (1960), Krawietz (1986), Holzapfel (2000), Haupt (2002), Bertram (2005), and Müller (2011).

The current position of any material point P (see Figure 3.1) in the three-dimensional space at time t is given by its position vector

$$\boldsymbol{x} = \chi\left(\boldsymbol{X}, t\right). \tag{3.1}$$

Material points are identified by their position vector $\boldsymbol{X}$ in the reference placement. The reference placement Ω_0 is defined by the body placement at $t = t_0$. Analogously, the current placement of the body at time $t \geq t_0$ is denoted by Ω_t. Point $\boldsymbol{x}$ lies in Ω_t (Truesdell and Toupin, 1960). The displacement $\boldsymbol{u}\left(\boldsymbol{X}, t\right)$ is the difference between the current and reference positions of a material point

$$\boldsymbol{u}(\boldsymbol{X}, t) = \boldsymbol{\chi}(\boldsymbol{X}, t) - \boldsymbol{X}. \tag{3.2}$$

Figure 3.1 shows the above-mentioned vectors for a specific material point P.

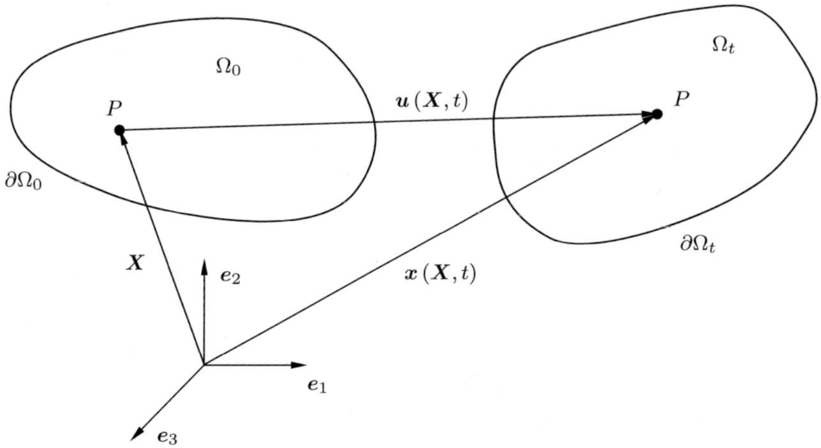

Figure 3.1: Reference placement Ω_0 and current placement Ω_t of point P

To highlight the separation towards large deformations, the Lagrangean and Eulerian description of a field quantity are introduced. The description of a field quantity ψ as a function of the reference position is called the Lagrangean description $\psi_{\mathrm{L}}(\boldsymbol{X}, t)$. The description of a field quantity ψ as a function of the current position is called the Eulerian description $\psi_{\mathrm{E}}(\boldsymbol{x}, t)$. The conversion between these two descriptions is performed with $\boldsymbol{\chi}$ and $\boldsymbol{\chi}^{-1}$

$$\psi_{\mathrm{E}}(\boldsymbol{x}, t) = \psi_{\mathrm{L}}\left(\boldsymbol{\chi}^{-1}(\boldsymbol{x}, t), t\right), \tag{3.3}$$

$$\psi_{\mathrm{L}}(\boldsymbol{X}, t) = \psi_{\mathrm{E}}(\boldsymbol{\chi}(\boldsymbol{X}, t), t). \tag{3.4}$$

Hereafter, the indices $(\cdot)_\mathrm{E}$ and $(\cdot)_\mathrm{L}$ are dropped and the formulations are differentiated solely by their arguments. The material time derivative is in the Lagrangean description given as (Truesdell and Toupin, 1960)

$$\dot{\psi}(\boldsymbol{X},t) = \frac{\partial\psi(\boldsymbol{X},t)}{\partial t}. \tag{3.5}$$

The material time derivative in the Eulerian description yields

$$\dot{\psi}(\boldsymbol{x},t) = \frac{\partial\psi(\boldsymbol{x},t)}{\partial t} + \frac{\partial\psi(\boldsymbol{x},t)}{\partial\boldsymbol{x}} \cdot \boldsymbol{v}(\boldsymbol{x},t). \tag{3.6}$$

Here, the velocity $\boldsymbol{v}(\boldsymbol{X},t) = \dot{\boldsymbol{x}}(\boldsymbol{X},t)$ is the material time derivative of $\boldsymbol{x}$. The deformation gradient $\boldsymbol{F}$ is defined as the gradient of the current position with respect to the reference position

$$\boldsymbol{F}(\boldsymbol{X},t) = \mathrm{Grad}(\boldsymbol{x}(\boldsymbol{X},t)) = \frac{\partial\boldsymbol{\chi}(\boldsymbol{X},t)}{\partial\boldsymbol{X}} = \frac{\partial\chi_i(\boldsymbol{X},t)}{\partial X_j}\boldsymbol{e}_i \otimes \boldsymbol{e}_j \tag{3.7}$$

In general, the deformation gradient is non-zero for pure rigid body rotations, even though the body does not deform. The displacement gradient is defined as

$$\boldsymbol{H}(\boldsymbol{X},t) = \mathrm{Grad}(\boldsymbol{u}(\boldsymbol{X},t)) = \frac{\partial\boldsymbol{u}(\boldsymbol{X},t)}{\partial\boldsymbol{X}} = \boldsymbol{F} - \boldsymbol{I}. \tag{3.8}$$

In this work, only small deformations are considered. This limitation is valid, if the following relation holds (Bertram, 2005)

$$\|\boldsymbol{H}\| = \sqrt{\mathrm{tr}(\boldsymbol{H}\boldsymbol{H}^\mathsf{T})} \ll 1. \tag{3.9}$$

The infinitesimal strain tensor is defined as follows (Bertram, 2005)

$$\boldsymbol{\varepsilon}(\boldsymbol{X},t) = \mathrm{sym}(\boldsymbol{H}(\boldsymbol{X},t)) = \frac{1}{2}\left(\boldsymbol{H}(\boldsymbol{X},t) + (\boldsymbol{H}(\boldsymbol{X},t))^\mathsf{T}\right). \tag{3.10}$$

A decomposition of the strain tensor ε into its spheric ε° and deviatoric part ε' is given by

$$\varepsilon = \varepsilon^\circ + \varepsilon'. \tag{3.11}$$

The spherical part describes the volume change (dilatation), and the deviatoric part describes the change of shape (distortion). These parts can be obtained with the two isotropic projectors $\mathbb{P}_1 = (\boldsymbol{I} \otimes \boldsymbol{I})/3$ and $\mathbb{P}_2 = \mathbb{I}^s - \mathbb{P}_1$

$$\varepsilon^\circ = \mathbb{P}_1[\varepsilon] = \frac{\mathrm{tr}(\varepsilon)}{3}\boldsymbol{I}, \quad \varepsilon' = \mathbb{P}_2[\varepsilon] = \varepsilon - \frac{\mathrm{tr}(\varepsilon)}{3}\boldsymbol{I}. \tag{3.12}$$

3.2 Balance equations

3.2.1 General formulation

Balance equations are crucial for predicting mechanical behavior. Therefore, the balance equation in a general form is introduced. It is postulated that the change of a volume integral of a field quantity ψ is equal to the sum of (i) the production p of ψ in volume Ω_t (ii) the supply s of ψ in volume Ω_t, and (iii) the flux $\boldsymbol{q}$ of ψ over the boundary $\partial\Omega_t$ (Truesdell and Toupin, 1960; Bertram, 2005)

$$\frac{\mathrm{d}}{\mathrm{d}t}\int_{\Omega_t} \psi \mathrm{d}v = \int_{\Omega_t} (p+s)\,\mathrm{d}v + \int_{\partial\Omega_t} \boldsymbol{q}\cdot \mathrm{d}\boldsymbol{a}. \tag{3.13}$$

Here, $\mathrm{d}\boldsymbol{a}$ is defined as $\mathrm{d}\boldsymbol{a} = \boldsymbol{n}\mathrm{d}a$, where $\mathrm{d}a$ is a surface element of the boundary $\partial\Omega_t$, and $\boldsymbol{n}$ is the outer surface normal unit vector. The application of the divergence theorem and the Reynolds transport theorem leads to the local formulation of the balance equation in regular points (Bertram, 2005)

$$\frac{\partial\psi}{\partial t} + \mathrm{div}(\psi\boldsymbol{v}) = p + s + \mathrm{div}(\boldsymbol{q}). \tag{3.14}$$

Considering the field quantity $\psi = \varrho$, the mass balance is obtained. Since the production p, supply s, and flux $\boldsymbol{q}$ are zero, the mass balance yields

$$\dot{\varrho} + \varrho \operatorname{div}(\boldsymbol{v}) = 0. \tag{3.15}$$

The change of mass can be calculated from $\operatorname{tr}(\boldsymbol{\varepsilon}) = (\mathrm{d}v - \mathrm{d}v_0)/\mathrm{d}v_0$ or $\varrho = \varrho_0 (1 - \operatorname{tr}(\boldsymbol{\varepsilon}))$. In the context of small deformations, the relation $\varrho \cong \varrho_0$ is an admissible approximation if $\|\boldsymbol{\varepsilon}\| \ll 1$ and all displacement fields fulfill the mass balance. Thus, the mass balance does not provide additional information in a displacement-based setting, but its validity is assumed later in the derivation of the linear momentum (Section 3.2.2).

3.2.2 Linear and angular momentum

The Cauchy stress tensor $\boldsymbol{\sigma}$ is a second order tensor describing the local stress state in a material point. The stress vector $\boldsymbol{t}$ on a surface with normal vector $\boldsymbol{n}$ can be obtained by Cauchy's Lemma

$$\boldsymbol{t} = \boldsymbol{\sigma}\boldsymbol{n}. \tag{3.16}$$

Similar to the strain tensor, the stress tensor can also be decomposed into a spherical part $\boldsymbol{\sigma}^\circ$ and a deviatoric part $\boldsymbol{\sigma}'$

$$\boldsymbol{\sigma} = \boldsymbol{\sigma}^\circ + \boldsymbol{\sigma}', \quad \boldsymbol{\sigma}^\circ = \mathbb{P}_1[\boldsymbol{\sigma}], \quad \boldsymbol{\sigma}' = \mathbb{P}_2[\boldsymbol{\sigma}]. \tag{3.17}$$

For hyperelastic materials, the fourth order stiffness tensor $\mathbb{C}$ maps the stress-inducing strain $\boldsymbol{\varepsilon} - \boldsymbol{\varepsilon}_{\mathrm{E}}$, that is the difference of the total strain $\boldsymbol{\varepsilon}$ and stress-free strain $\boldsymbol{\varepsilon}_{\mathrm{E}}$, linearly onto the stress $\boldsymbol{\sigma}$

$$\boldsymbol{\sigma} = \mathbb{C}[\boldsymbol{\varepsilon} - \boldsymbol{\varepsilon}_{\mathrm{E}}]. \tag{3.18}$$

The existence of an elastic energy and the symmetry of second derivatives imply the main symmetry of the stiffness tensor $\mathbb{C} = \mathbb{C}^{\mathsf{T}_\mathsf{M}}$. The main transposition of a fourth order tensor describes the exchange of the left and right index pair ($C^{\mathsf{T}_\mathsf{M}}_{ijkl} = C_{klij}$). The symmetry of the strain tensor implies a right sub-symmetry of the stiffness tensor $\mathbb{C} = \mathbb{C}^{\mathsf{T}_\mathsf{R}}$. The right transposition of a fourth order tensor describes the exchange of the left and two indices ($C^{\mathsf{T}_\mathsf{R}}_{ijkl} = C_{jikl}$). As a result, the stiffness tensor also has the left sub-symmetry $\mathbb{C} = \mathbb{C}^{\mathsf{T}_\mathsf{L}}$. The left transposition is defined analogously ($C^{\mathsf{T}_\mathsf{L}}_{ijkl} = C_{ijlk}$).

To derive the linear momentum balance, the momentum density $\varrho \boldsymbol{v}$, a first order tensor, is balanced. The general form of a balance equation of tensorial quantities is analogous to the scalar form (Equation (3.14)). The production is zero $\mathbf{0}$, the supply is $\varrho \boldsymbol{b}$, and the flux linear momentum is $\boldsymbol{\sigma}$. The local formulation of the balance of linear momentum under consideration of the mass balance (Equation (3.15)) yields the following (Truesdell and Toupin, 1960)

$$\varrho \dot{\boldsymbol{v}} = \varrho \boldsymbol{b} + \operatorname{div}(\boldsymbol{\sigma})\,. \tag{3.19}$$

The formulation simplifies in the quasi-static case to

$$\mathbf{0} = \varrho \boldsymbol{b} + \operatorname{div}(\boldsymbol{\sigma})\,. \tag{3.20}$$

If moments densities are absent and a Boltzmann continuum is considered, the quasi-static balance of angular momentum results in the symmetry of the stress tensor (Truesdell and Toupin, 1960)

$$\boldsymbol{\sigma} = \boldsymbol{\sigma}^{\mathsf{T}}. \tag{3.21}$$

3.2.3 Energy and entropy

Because the kinetic energy balance is obtained by multiplying the impulse balance (Equation (3.19)) with the velocity $\boldsymbol{v}$, it does not represent an independent equation (Truesdell and Toupin, 1960)

$$\frac{1}{2}\varrho\left(\boldsymbol{v}\cdot\boldsymbol{v}\right)^{\cdot}=\varrho\boldsymbol{b}\cdot\boldsymbol{v}+\operatorname{div}\left(\boldsymbol{\sigma}^{\mathsf{T}}\boldsymbol{v}\right)-\boldsymbol{\sigma}\cdot\dot{\boldsymbol{\varepsilon}}. \tag{3.22}$$

Here, the power of external forces $\operatorname{div}\left(\boldsymbol{\sigma}^{\mathsf{T}}\boldsymbol{v}\right)=\operatorname{div}\left(\boldsymbol{\sigma}\boldsymbol{v}\right)$ is the flux of kinetic energy over the body surface, $\varrho\boldsymbol{b}\cdot\boldsymbol{v}$ is the supply of kinetic energy, and $-\boldsymbol{\sigma}\cdot\dot{\boldsymbol{\varepsilon}}$ is the production of kinetic energy given by the stress power.
The total volume-specific energy equals the sum of the volume-specific internal energy ϱe and volume-specific kinetic energy $\varrho\boldsymbol{v}\cdot\boldsymbol{v}/2$. The local form of the balance of total energy in regular points is therefore (Truesdell and Toupin, 1960)

$$\varrho\dot{e}+\frac{1}{2}\varrho\left(\boldsymbol{v}\cdot\boldsymbol{v}\right)^{\cdot}=\varrho w-\operatorname{div}\left(\boldsymbol{q}\right)+\varrho\boldsymbol{b}\cdot\boldsymbol{v}+\operatorname{div}\left(\boldsymbol{\sigma}\boldsymbol{v}\right). \tag{3.23}$$

The subtraction of the mechanical energy from the balance of total energy yields the balance of the internal or non-mechanical energy

$$\varrho\dot{e}=\varrho w+\boldsymbol{\sigma}\cdot\dot{\boldsymbol{\varepsilon}}-\operatorname{div}\left(\boldsymbol{q}\right) \tag{3.24}$$

where $\boldsymbol{\sigma}\cdot\boldsymbol{\varepsilon}+\varrho w$ represents the source term. Vector $\boldsymbol{q}$ describes the non-convective heat flux, that is, the flux of internal energy through the material.
The entropy balance can be stated as (Coleman and Noll, 1963)

$$\varrho\dot{\eta}=\frac{\varrho w}{\theta}-\operatorname{div}\left(\frac{\boldsymbol{q}}{\theta}\right)+\varrho p_{\eta}. \tag{3.25}$$

Here, η is the mass-specific entropy, $\varrho w/\theta$ is the supply of entropy, p_{η} is the entropy production, and $\boldsymbol{q}/\theta$ is the entropy flux. The second law of thermodynamics states that, in all thermo-kinetic processes, the entropy production is non-negative

$$p_\eta \geq 0. \tag{3.26}$$

3.3 Composite micromechanics

3.3.1 Fundamentals

As you recall, Figure 2.3 presented both the component scale and some microstructure scales. A separation of the microstructure and component scales is not admissible, however, and so an representative volume element (RVE) does not exist. Traditional homogenization schemes are based on the assumption of a scale separation and the existence of an RVE. The application of mean-field homogenization schemes in SMC is state of the art and leads to satisfying results in many two-scale structural simulations (see Section 1.2). Nevertheless, the author regards the missing scale separation as a questionable assumption. One modern homogenization framework that does not require scale separation is the theory of computational continua introduced by Fish and Kuznetsov (2010) and Fish et al. (2015). Ostoja-Starzewski (2002; 2006) discussed in detail the missing separation between the microscale and macroscale with the statistical volume element and apparent properties.

For a more detailed introduction to microstructural mechanics, the reader is referred to the textbooks of Nemat-Nasser and Hori (1999), Torquato (2002), and Gross and Seelig (2011).

In a heterogeneous microstructure, stress and strain fields are also generally microheterogeneous. They are given by $\boldsymbol{\varepsilon}(\boldsymbol{x})$ and $\boldsymbol{\sigma}(\boldsymbol{x})$. If cracks and pores are absent the effective macroscopic properties, denoted by an overbar $\overline{(\cdot)}$, are obtained by taking a volume average (see Section 2.2) within an RVE

$$\bar{\varepsilon} = \langle \varepsilon (\boldsymbol{x}) \rangle, \quad \bar{\boldsymbol{\sigma}} = \langle \boldsymbol{\sigma} (\boldsymbol{x}) \rangle. \tag{3.27}$$

In case of linear elasticity, the relation between the effective stress $\bar{\boldsymbol{\sigma}}$ and effective strain $\bar{\varepsilon}$ is given by the thus implicitly defined effective stiffness $\bar{\mathbb{C}}$

$$\bar{\boldsymbol{\sigma}} = \bar{\mathbb{C}} [\bar{\varepsilon}] . \tag{3.28}$$

If linear elasticity in each phase is assumed and absence of stress-free strains, the strain localization tensor $\mathbb{A} (\boldsymbol{x})$ describes the relation between the local and the effective strain

$$\varepsilon (\boldsymbol{x}) = \mathbb{A} (\boldsymbol{x}) [\bar{\varepsilon}] . \tag{3.29}$$

Analogously, the stress localization tensor $\mathbb{B} (\boldsymbol{x})$ describes the relation between the local and the effective stress, if eigenstresses are absent

$$\boldsymbol{\sigma} (\boldsymbol{x}) = \mathbb{B} (\boldsymbol{x}) [\bar{\boldsymbol{\sigma}}] . \tag{3.30}$$

The effective stiffness can thus be obtained as follows

$$\bar{\mathbb{C}} = \langle \mathbb{C} (\boldsymbol{x}) \mathbb{A} (\boldsymbol{x}) \rangle = \langle (\mathbb{C} (\boldsymbol{x}))^{-1} \mathbb{B} (\boldsymbol{x}) \rangle^{-1}. \tag{3.31}$$

Hill (1963) stated that the macroscopic stress power density and the phase average of the microscopic stress power density are equivalent if at least one of the following conditions is fulfilled:

- any volume subjected to homogeneous Dirichlet boundary conditions
- any volume subjected to homogeneous Neumann boundary conditions
- any volume subjected to periodic boundary conditions
- in an ergodic volume at the infinite volume limit.

The Hill condition for a divergence free stress field $\boldsymbol{\sigma}$ and a compatible strain rate field $\dot{\varepsilon}$ yields

$$\langle \boldsymbol{\sigma}(\boldsymbol{x}) \cdot \dot{\varepsilon}(\boldsymbol{x}) \rangle = \langle \boldsymbol{\sigma}(\boldsymbol{x}) \rangle \cdot \langle \dot{\varepsilon}(\boldsymbol{x}) \rangle = \bar{\boldsymbol{\sigma}} \cdot \dot{\bar{\varepsilon}}. \tag{3.32}$$

The strain rate and stress fields can be split into an average $\bar{(\cdot)}$ part and a fluctuating $\hat{(\cdot)}$ part

$$\boldsymbol{\sigma}(\boldsymbol{x}) = \bar{\boldsymbol{\sigma}} + \hat{\boldsymbol{\sigma}}(\boldsymbol{x}), \tag{3.33}$$

$$\dot{\varepsilon}(\boldsymbol{x}) = \dot{\bar{\varepsilon}} + \dot{\hat{\varepsilon}}(\boldsymbol{x}). \tag{3.34}$$

The Hill condition implies that the stress and strain rate fluctuations are orthogonal, on average

$$\langle \hat{\boldsymbol{\sigma}}(\boldsymbol{x}) \cdot \dot{\hat{\varepsilon}}(\boldsymbol{x}) \rangle = 0. \tag{3.35}$$

In the special case of elasticity, the Hill condition can be rewritten in terms of the elastic energy $\varepsilon(\boldsymbol{x}) \cdot \boldsymbol{\sigma}(\boldsymbol{x})$

$$\langle \boldsymbol{\sigma}(\boldsymbol{x}) \cdot \varepsilon(\boldsymbol{x}) \rangle = \langle \boldsymbol{\sigma}(\boldsymbol{x}) \rangle \cdot \langle \varepsilon(\boldsymbol{x}) \rangle = \bar{\boldsymbol{\sigma}} \cdot \bar{\varepsilon}. \tag{3.36}$$

3.3.2 Mori-Tanaka homogenization scheme

As discussed in Section 1.2, in contrast to bounding homogenization schemes, the Mori-Tanaka homogenization scheme is an estimating mean-field homogenization procedure. Originally introduced by Mori and Tanaka (1973), their method has been applied in homogenization procedures for a wide range of heterogeneous materials. In this Section, an expression for the effective stiffness is derived briefly followed by a discussion of the advantages and, especially, the drawbacks of this method. For a more detailed description, the reader is referred to the representations of Ponte Castañeda and Willis (1995), Torquato (2002), and Brylka (2017).

The key assumption of the Mori-Tanaka homogenization scheme is that the inclusion strain is estimated by the strain localization of the single inclusion problem (SIP) and the average matrix strain

$$\varepsilon_{\mathrm{F}} = \langle \mathbb{A}_{\mathrm{F0}}^{\mathrm{SIP}} \rangle_{\mathrm{F}} \left[\varepsilon_{\mathrm{M}} \right] = \langle \mathbb{A}_{\mathrm{F0}}^{\mathrm{SIP}} \rangle_{\mathrm{F}} \mathbb{A}_{\mathrm{M}}^{\mathrm{MT}} \left[\bar{\varepsilon} \right] = \mathbb{A}_{\mathrm{F}}^{\mathrm{MT}} \left[\bar{\varepsilon} \right] . \tag{3.37}$$

The operator $\langle \cdot \rangle_{\mathrm{F}}$ is defined as the volume average in the fiber volume. Analogously, the operator $\langle \cdot \rangle_{\mathrm{M}}$ is defined as the volume average in the matrix volume. Eshelby (1957; 1959) described the elastic solution of an ellipsoidal inclusion in an infinite matrix (SIP), which allows one to explicitly express the SIP strain localization tensor in a reference orientation $\langle \cdot \rangle_0$

$$\mathbb{A}_{\mathrm{F0}}^{\mathrm{SIP}} = \left(\mathbb{I}^{\mathrm{s}} + \mathbb{P}_0 \left(\mathbb{C}_{\mathrm{F}} - \mathbb{C}_{\mathrm{M}} \right) \right)^{-1}. \tag{3.38}$$

An expression for the symmetric, transversely-isotropic polarization tensor $\mathbb{P}_0 = \mathbb{E}_0 \mathbb{C}_{\mathrm{M}}^{-1}$ can be found in Ponte Castañeda and Suquet (1998) or Torquato (2002). Here, $\mathbb{E}_0$ is the Eshelby tensor. In order to maintain a compact scope of this work, an extensive discussion of the eigenstrain concept and the Eshelby solution is avoided. The average over the total fiber volume $\langle \cdot \rangle_{\mathrm{F}}$ is equivalent to a fiber orientation average. For a computationally efficient orientation average of symmetric, transversely-isotropic tensors, the reader is referred to the work of Advani and Tucker III (1987).

The relation $\mathbb{I}^{\mathrm{s}} = c_{\mathrm{F}} \mathbb{A}_{\mathrm{F}}^{\mathrm{MT}} + c_{\mathrm{M}} \mathbb{A}_{\mathrm{M}}^{\mathrm{MT}}$ yields the phase-averaged strain localization tensors in the matrix and the fiber phase

$$\mathbb{A}_{\mathrm{M}}^{\mathrm{MT}} = \left(c_{\mathrm{F}} \langle \mathbb{A}_{\mathrm{F0}}^{\mathrm{SIP}} \rangle_{\mathrm{F}} + c_{\mathrm{M}} \mathbb{I}^{\mathrm{s}} \right)^{-1}, \tag{3.39}$$

$$\mathbb{A}_{\mathrm{F}}^{\mathrm{MT}} = \frac{1}{c_{\mathrm{F}}} \left(\mathbb{I}^{\mathrm{s}} - c_{\mathrm{M}} \left(c_{\mathrm{F}} \langle \mathbb{A}_{\mathrm{F0}}^{\mathrm{SIP}} \rangle_{\mathrm{F}} + c_{\mathrm{M}} \mathbb{I}^{\mathrm{s}} \right)^{-1} \right). \tag{3.40}$$

Since $\mathbb{A}_{\mathrm{M}}^{\mathrm{MT}}$ and $\mathbb{A}_{\mathrm{F}}^{\mathrm{MT}}$ are known, the calculation of the macroscopic stiffness $\mathbb{C}^{\mathrm{MT}}$ is straight-forward

$$\bar{\mathbb{C}}^{\mathrm{MT}} = \langle \mathbb{C}\mathbb{A} \rangle \tag{3.41}$$

$$= \mathbb{C}_{\mathrm{M}} + c_{\mathrm{F}} \langle (\mathbb{C}_{\mathrm{F}} - \mathbb{C}_{\mathrm{M}})\, \mathbb{A}_{\mathrm{F0}}^{\mathrm{SIP}} \rangle_{\mathrm{F}} \mathbb{A}_{\mathrm{M}}^{\mathrm{MT}} \tag{3.42}$$

$$= \mathbb{C}_{\mathrm{M}} + c_{\mathrm{M}} \langle (\mathbb{C}_{\mathrm{F}} - \mathbb{C}_{\mathrm{M}})\, \mathbb{A}_{\mathrm{F0}}^{\mathrm{SIP}} \rangle_{\mathrm{F}} \left(c_{\mathrm{F}} \langle \mathbb{A}_{\mathrm{F0}}^{\mathrm{SIP}} \rangle_{\mathrm{F}} + c_{\mathrm{M}} \mathbb{I}^{\mathrm{s}} \right)^{-1}. \tag{3.43}$$

In this widespread formulation (Benveniste, 1987; Norris, 1989; Benveniste et al., 1991), the main symmetry of the effective stiffness tensor $\bar{\mathbb{C}}^{\mathrm{MT}}$ cannot be ensured. Therefore, in the following, isotropy of the fiber stiffness $\mathbb{C}_{\mathrm{F}}$ and the matrix stiffness $\mathbb{C}_{\mathrm{M}}$ is assumed. Equation (3.38) and further simplifications thus lead to

$$\bar{\mathbb{C}}^{\mathrm{MT}} = \mathbb{C}_{\mathrm{M}} + c_{\mathrm{F}} \left(c_{\mathrm{M}} \left\langle \left(\mathbb{P}_0 + (\mathbb{C}_{\mathrm{F}} - \mathbb{C}_{\mathrm{M}})^{-1} \right)^{-1} \right\rangle_{\mathrm{F}}^{-1} + c_{\mathrm{F}} \left(\mathbb{C}_{\mathrm{F}} - \mathbb{C}_{\mathrm{M}} \right) \right)^{-1}. \tag{3.44}$$

Since the inversion and the orientation average maintain the symmetry of a tensor, Equation (3.44) represents a proof of the symmetry of the effective stiffness tensor, in the case of inclusions of the same type and elastically isotropic phases.

The advantages of the Mori-Tanaka scheme are the short and clear derivation. Its application has been extensively validated and is widely accepted. A numerical implementation is compact, with low computational effort.

There is no variational formulation of the Mori-Tanaka estimate including a valid elastic energy which would, for example, ensure the symmetry of the stiffness tensor. This is a deficit of the Mori-Tanaka estimate (Ponte Castañeda and Willis, 1995). Benveniste et al. (1991) showed that, in general, the Mori-Tanaka estimate leads to an asymmetric stiffness with more than one inclusion phase of non-similar geometry. However, they also showed that, for the case of special coated fibers in a matrix, the symmetry can be ensured (Benveniste et al., 1989). Ponte Castañeda and Willis (1995) stated that, for anisotropic constituents, the symmetry of the stiffness tensor is only ensured if the inclusion shape is isotropic (spheroidal). Ponte Castañeda and Willis (1995) provide an example of spherical inclusions distributed anisotropically, assuming isotropy of the matrix and inclusion phases, which leads to an isotropic macroscopic stiffness. The Mori-Tanaka stiffness may violate bounds of the Hashin-Shtrikman type, for example, with materials having isotropic distributions of randomly oriented cracks and disks (Ponte Castañeda and Willis, 1995), or with three phase materials having randomly oriented ellipsoids (Norris, 1989). Ferrari (1991) showed that, for high inclusion volume fractions, the Mori-Tanka estimate may lead to unphysical results, such as a dependence of the effective stiffness on the matrix stiffness at the vanishing matrix volume limit.

Chapter 4

Biaxial tensile tests and microstructure-based inverse parameter identification of macroscopically inhomogeneous SMC[1]

4.1 Introduction

Due to their good formability, economical mass-production and high specific strength and stiffness, discontinuous fiber-reinforced polymers are increasingly applied in components in the automotive industry and the mobility sector in general. The application of this class of materials is, however, hindered by a lack of a detailed understanding of its mechanical behavior.

The material class under consideration is the discontinuous fiber-reinforced thermoset, sheet molding compound (SMC). Specifically, glass fiber-reinforced vinylester (VE) resin is under consideration. The fibers have a length of $25\mathrm{mm}$. Typically, parts of SMC such as structural automotive components are manufactured by compression molding of

[1] This chapter is based on the paper "Biaxial tensile tests and microstructure-based inverse parameter identification of inhomogeneous SMC composites" (Schemmann et al., 2018a).

prepregs. This forming process leads to a macroscopically heterogeneous fiber orientation distribution. As a result, the material behavior is spatially heterogeneous and anisotropic in a process sensitive way (Katayama et al., 2001).

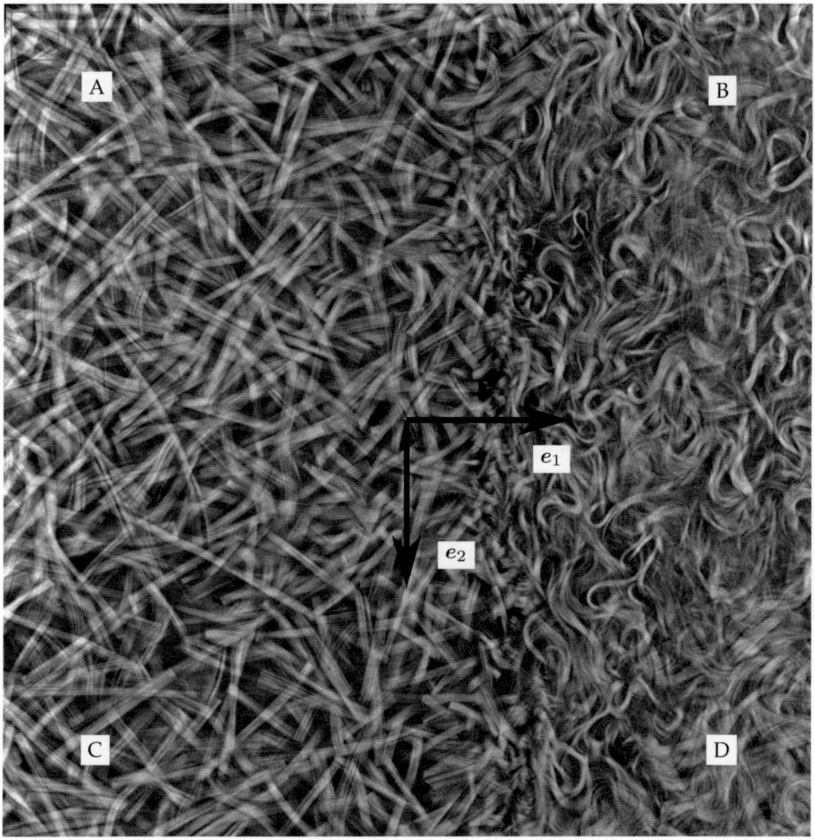

Figure 4.1: μCT scan of a macro-heterogeneous SMC microstructure that shows to different flow regions: almost straight fibers on the left side and curly fibers on the right third of the image, specimen dimensions: $73\text{mm} \times 73\text{mm} \times 3\text{mm}$ (Pinter, 2017), fourth order fiber orientation tensors in the marked points are listed in Appendix A

In this study, we aimed for samples that showed a significant macro-heterogeneity on the scale of the cruciform specimen. The here considered samples were molded by Hohberg et al. (2017b) with an approximate initial mold coverage of 20%. An example for such a heterogeneous microstructure is shown in a slice of a micro-computed tomography (μCT) scan in Figure 4.1. The here shown heterogeneities are more pronounced than typically observed in SMC. A typical SMC microstructure is depicted in Figure 2.5 and Figure 6.1.

Pinter et al. (2018) applied fiber orientation analysis to estimate the fiber orientation distributions in this images. Figure 4.2 depicts the spatial distribution of the first (highest) eigenvalue of the second order fiber orientation tensor of the first kind $\boldsymbol{N}$. Arrows indicate the in-plane projection of the corresponding eigenvector. Figure 4.3 and Figure 4.4 visualize the second and third eigenvectors accordingly.

Within the virtual process chain, this heterogeneity has been taken into account in moldfilling analysis (Advani and Tucker III, 1987; Osswald and Tucker, 1989) and homogenization approaches (Buck et al., 2015; Müller et al., 2015b) for discontinuous fiber-reinforced polymers. The mechanical characterization of heterogeneous composites has been considered in great detail in the literature by the assumption of homogeneously distributed material properties within the specimen (e.g. by Zhou et al. (2017); Trauth et al. (2018)). However, the consideration of heterogeneous specimens is in an early stage of research.

4.2 Experimental procedure

The specimens are tested in an electro-mechanical biaxial tensile testing machine from Zwick. Four independent axes allow for a maximum load of 150kN. Figure 5.1a (left) shows the setup of the machine. Five markers for integrated strain measurements are located on the lower side of the specimen.

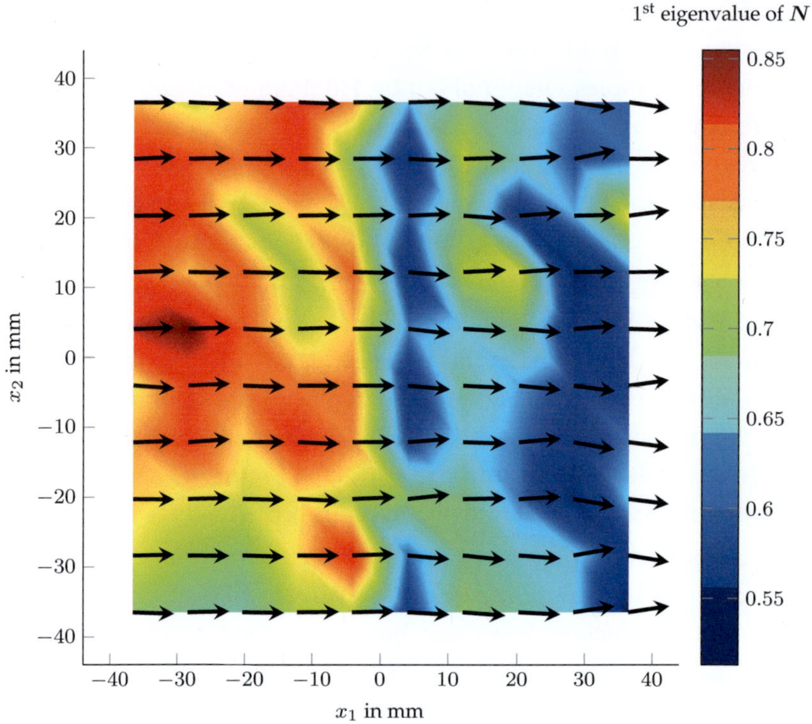

Figure 4.2: First eigenvalue of the second order fiber orientation tensor $\boldsymbol{N}$ and the in-plane projection of the corresponding eigenvector

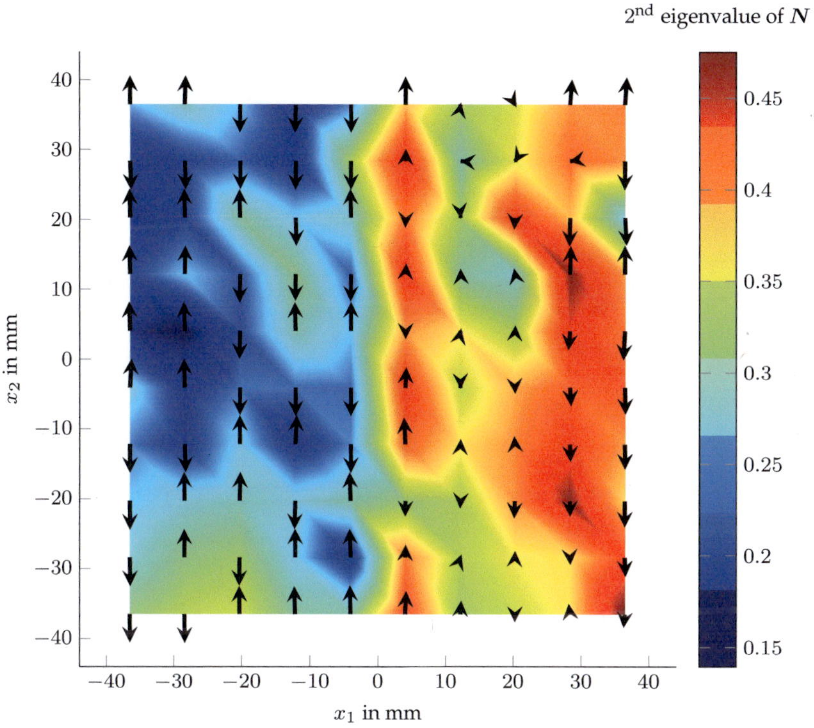

Figure 4.3: Second eigenvalue of the second order fiber orientation tensor $\boldsymbol{N}$ and the in-plane projection of the corresponding eigenvector

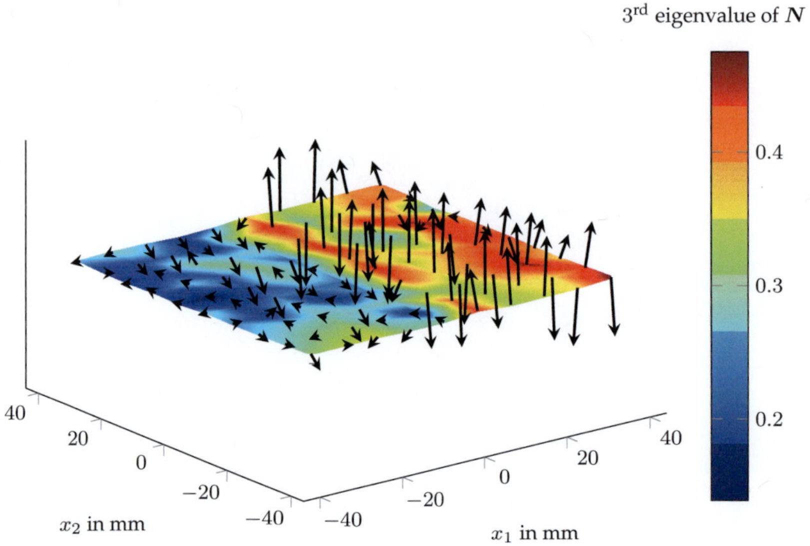

Figure 4.4: Third eigenvalue of the second order fiber orientation tensor $\boldsymbol{N}$ and the the corresponding eigenvector

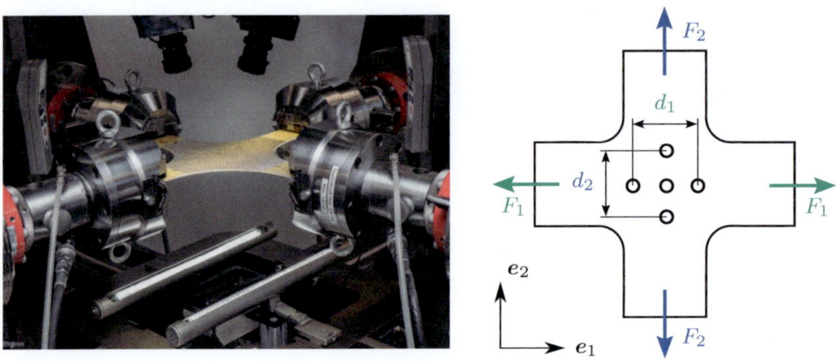

Figure 4.5: Biaxial tensile testing machine (left) and cruciform specimen with markers for displacement measurement (right)

These markers are traced by the integrated strain measurement system Video XTens. The distances between opposing points d_1 and d_2 (see Figure 5.1a (right)) are used for the load control. Additionally, the marker positions are used for the active midpoint control of the system which allows for load application without bending in specimen's arms, even for highly heterogeneous specimen. On the upper side of the specimen, a speckle pattern is applied which is used for full-field strain measurements with the GOM ARAMIS 4M grayscale image correlation system.

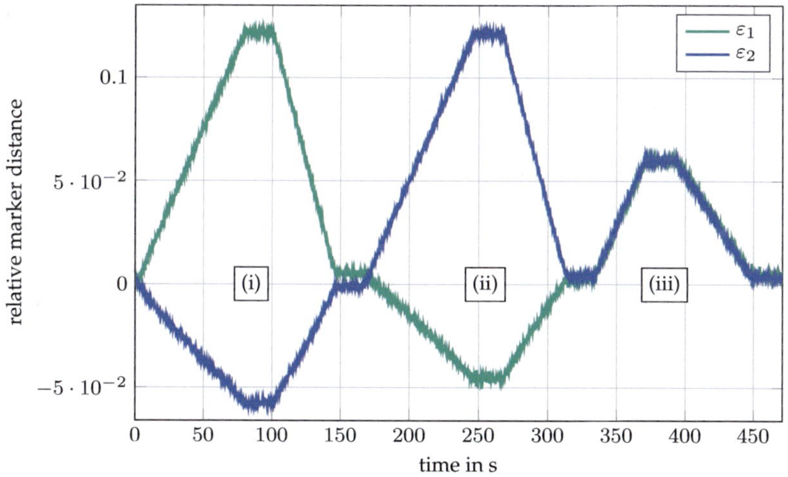

Figure 4.6: Experimental procedure: strain path

Figure 4.6 shows the prescribed strain path, and Figure 4.7 the force path over the three loading steps of the experiments. Firstly (i), the specimen is loaded with a relative marker distance or average technical strain ε_1. The perpendicular axes are in the e_2-direction and controlled in such a way that the force remains at a low contact force. This strain is maintained whereas the viscosity of the material leads to a decrease of the tension forces. Secondly (ii), after a relaxation time, the e_2-direction is tested equivalently. Thirdly (iii), an equibiaxial load is applied to the specimen. The clearly visible scattering in Figure 4.6 is likely to derive from the high measuring rate and vibrations of the sensor mounting. The relative marker distance is not considered in the experimental evaluation.

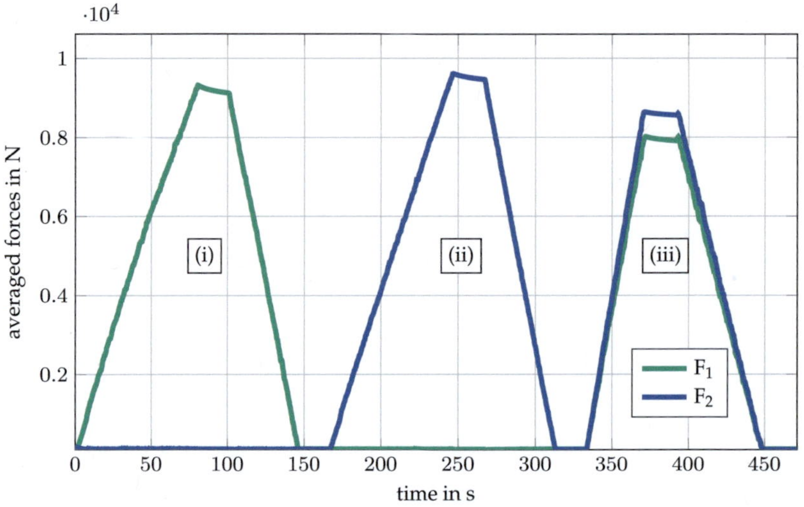

Figure 4.7: Experimental procedure: force path

4.3 Inverse parameter identification

In biaxial tensile experiments, the stress and strain fields in the specimen are heterogeneous. Due to this heterogeneity, the material parameters cannot be identified directly, but may be obtained by the solution of an inverse problem. The direct problem is defined by the strain displacement relation, the balance of linear momentum, the boundary conditions and Hooke's law for the constitutive relation between the stress tensor $\boldsymbol{\sigma}$ (with $\boldsymbol{\sigma} = \boldsymbol{\sigma}^{\mathsf{T}}$) and strain tensor $\boldsymbol{\varepsilon}$ with a position-dependent stiffness $\mathbb{C}(\boldsymbol{x})$

$$\boldsymbol{\sigma} = \mathbb{C}\,(\boldsymbol{x})\,[\boldsymbol{\varepsilon}]\,. \tag{4.1}$$

The quasi-static boundary value problem - using these relations - is solved with the FE software ABAQUS/Standard.

Furthermore, if the numerically calculated strain field is $\boldsymbol{\varepsilon}^{\text{sim}} \in \mathcal{U}^{\text{sim}}$, whereas $\mathcal{C}$ is the space of stiffness tensors and $\mathcal{U}^{\text{sim}}$ the solution space of the FEM for strains, the inverse problem can be defined by:

$$\text{find } \mathbb{C}(\boldsymbol{x}) \in \mathcal{C}, \text{ that resolves } \boldsymbol{\varepsilon}^{\text{sim}}\left(\mathbb{C}(\boldsymbol{x})\right) = \boldsymbol{\varepsilon}^{\text{exp}}. \tag{4.2}$$

In general, the equality of the strain fields $\boldsymbol{\varepsilon}^{\text{exp}}$ and $\boldsymbol{\varepsilon}^{\text{sim}}(\mathbb{C}(\boldsymbol{x}))$ cannot be assured. This discrepancy derives from measurement uncertainties or modeling errors (Mahnken and Stein, 1996b). A solution of the inverse problem is, however, obtained by minimizing the correlation function $f(\mathbb{C}(\boldsymbol{x}))$ that describes the difference of the experimental and simulated strains. The vector $\underline{r}(\mathbb{C}(\boldsymbol{x}))$ quantifies the deviation of the strain fields. In $f(\mathbb{C}(\boldsymbol{x}))$ and $\underline{r}(\mathbb{C}(\boldsymbol{x}))$, only the measurable in-plane strains are considered

$$\underline{r}(\mathbb{C}(\boldsymbol{x})) = \begin{pmatrix} \varepsilon^{\text{sim}}_{11,i,j}(\mathbb{C}(\boldsymbol{x})) - \varepsilon^{\text{exp}}_{11,i,j} \\ \varepsilon^{\text{sim}}_{22,i,j}(\mathbb{C}(\boldsymbol{x})) - \varepsilon^{\text{exp}}_{22,i,j} \\ \gamma^{\text{sim}}_{12,i,j}(\mathbb{C}(\boldsymbol{x})) - \gamma^{\text{exp}}_{12,i,j} \\ \vdots \end{pmatrix}_{i=1\ldots K, j=1\ldots N} . \tag{4.3}$$

Since the meshes of the experimentally measured strain field is discretized much finer than the FEM simulation solution, both strain fields are compared in a comparison mesh. A mapping algorithm interpolates the strain fields on a comparison mesh with the spatial coordinates $\boldsymbol{x}_i \in \Omega,\ i = 1...K$ and discrete times $t_j \in \mathcal{T},\ j = 1...N$.

Using the strain vector $\underline{\varepsilon} = (\varepsilon_{11}, \varepsilon_{22}, \gamma_{12})^{\mathsf{T}}$, the error function $\underline{f}(\mathbb{C}(\boldsymbol{x}))$ and the strain deviation $\underline{r}(\mathbb{C}(\boldsymbol{x}))$, the inverse problem can be rewritten as follows

$$f(\mathbb{C}(\boldsymbol{x})) = ||\underline{r}(\mathbb{C}(\boldsymbol{x}))||_2^2 = \underline{r}(\mathbb{C}(\boldsymbol{x}))^{\mathsf{T}}\underline{r}(\mathbb{C}(\boldsymbol{x})) \rightarrow \min_{\mathbb{C}(\boldsymbol{x})\in C} . \tag{4.4}$$

The results of experiments with the identification of a homogeneous anisotropic stiffness and consideration of viscous effects in SMC and long fiber-reinforced thermoplastics can be found in Schemmann et al. (2015a). These results are compared with homogenized material over the whole specimen with the interaction direct derivative (IDD) estimate (Schemmann et al., 2015b).

4.3.1 Implications by the assumption of linearity between the stiffness tensor and the fiber orientation tensor

The harmonic decomposition of the fully anisotropic stiffness tensor (Schouten, 1924; Forte and Vianello, 1996; Rychlewski, 2000; Böhlke et al., 2010) yields

$$\underbrace{\mathbb{C}}_{21} = \underbrace{\overbrace{\underbrace{k_1\mathbb{P}_1 + k_2\mathbb{P}_2}_{2}}^{\mathbb{C}_{\mathrm{iso}}} + \overbrace{\underbrace{\boldsymbol{I}\otimes\boldsymbol{H}_1' + \boldsymbol{H}_1'\otimes\boldsymbol{I}}_{5}}^{\mathbb{C}_{\mathrm{H21}}} + \overbrace{\underbrace{4\mathbb{J}(\boldsymbol{H}_2')}_{5}}^{\mathbb{C}_{\mathrm{H22}}} + \overbrace{\underbrace{\mathbb{H}'}_{9}}^{\mathbb{C}_{\mathrm{H4}}}}_{21}, \tag{4.5}$$

where the operator $\mathbb{J}$ is defined as

$$4\mathbb{J}[\boldsymbol{A}] = (A_{im}\delta_{jn} + A_{in}\delta_{jm} + A_{jn}\delta_{im} + A_{jm}\delta_{in})\,\boldsymbol{e}_i \otimes \boldsymbol{e}_j \otimes \boldsymbol{e}_m \otimes \boldsymbol{e}_n . \tag{4.6}$$

The two isotropic projectors are $\mathbb{P}_1 = (\boldsymbol{I}\otimes\boldsymbol{I})/3$ and $\mathbb{P}_2 = \mathbb{I}^S - \mathbb{P}_1$, $\mathbb{I}^S$ denotes the fourth-order identity on symmetric second-order tensors. The tensors $\boldsymbol{H}_1'$, $\boldsymbol{H}_2'$ and $\mathbb{H}'$ are completely symmetric and traceless. The coefficients h_1 and h_2 assemble the isotropic part $\mathbb{C}_{\text{iso}}$, $\mathbb{C}_{\text{H21}}$ and $\mathbb{C}_{\text{H22}}$ are first and second deviatoric parts, $\mathbb{C}_{\text{H4}}$ is the harmonic part. The numbers below the quantities, i.e. in Equation (4.5), indicate the number of independent components of each tensor. The assumption of linearity between $\mathbb{C}$ and the fiber orientation tensors of the third kind $\boldsymbol{D}', \mathbb{D}'$ (see Section 2.4.3) lead to the following relationship:

$$\mathbb{C} = \underbrace{k_1\mathbb{P}_1 + k_2\mathbb{P}_2}_{2} \quad + \underbrace{\alpha(\boldsymbol{I} \otimes \boldsymbol{D}' + \boldsymbol{D}' \otimes \boldsymbol{I})}_{1} \quad + \underbrace{\beta\mathbb{J}(\boldsymbol{D}')}_{1} \quad + \underbrace{\gamma\mathbb{D}'}_{1}. \tag{4.7}$$

If $\mathbb{D}'$ and $\boldsymbol{D}'$ are known, this assumption reduces the number of unknown parameters from 21 (anisotropic elasticity) to five, namely k_1, k_2, α, β and γ. These parameters depend on the matrix and the fiber properties, the volume fraction, the fiber geometry, but not the fiber orientation distribution, since this is explicitly taken into account by the orientation tensors. Therefore, we assume these parameters as constant within the specimen. The material inhomogeneity derives only from an inhomogeneous fiber orientation distribution which can be obtained, i.e. by μCT analysis or moldfilling simulations.

4.3.2 Numerical treatment of the optimization

The parameter k_1 describes the relation between the stress and the volume change or trace of the strain $\text{tr}(\boldsymbol{\varepsilon}) = \varepsilon_{11} + \varepsilon_{22} + \varepsilon_{33}$ and ε_{33}. Since the out of plane strain is not measurable (Equation (4.3)), k_1 is indifferent of the objective function. If all the other parameters are combined into the parameter vector $\underline{p} = (k_2, \alpha, \beta, \gamma)$ the stiffness can be written as

$$\mathbb{C}(\boldsymbol{x}) = \mathbb{C}(\underline{p}, k_1, \boldsymbol{D}'(\boldsymbol{x}), \mathbb{D}'(\boldsymbol{x})) = \mathbb{C}(\underline{p}). \tag{4.8}$$

A reformulation of the optimization problem leads to

$$f(\underline{p}) = \underline{r}(\underline{p})^{\mathsf{T}} \underline{r}(\underline{p}) \to \min_{\underline{p} \in \mathcal{P}} . \tag{4.9}$$

The given problem is nonlinear and solved numerically. To identify the unknown parameters $\underline{p}$, a Gauss-Newton optimization procedure is used Lecompte et al. (2007). This procedure is based on the classical Newton's optimization method

$$\underline{p}^{k+1} = \underline{p}^{k} - \left[\underline{\underline{H}}(\underline{p}^{k})\right]^{-1} \nabla f(\underline{p}^{k}). \tag{4.10}$$

The gradient of Equation (4.9) can be written as

$$\nabla f(\underline{p}) = 2\underline{\underline{J}}(\underline{p})^{\mathsf{T}} \underline{r}(\underline{p}), \text{ with the Jacobian } J_{ij}(\underline{p}) = \frac{\partial r_i(\underline{p})}{\partial p_j}. \tag{4.11}$$

The Gauss-Newton-Approximation of the Hessian can than be expressed as follows

$$\underline{\underline{H}}(\underline{p}) = 2\underline{\underline{J}}(\underline{p})^{\mathsf{T}} \underline{\underline{J}}(\underline{p}) + 2\underline{\underline{G}}(\underline{p}), \text{ with } G_{ij}(\underline{p}) = \sum_{k=1}^{N \cdot K} \frac{\partial^2 r_k(\underline{p})}{\partial p_i \partial p_j} r_k(\underline{p}). \tag{4.12}$$

Negligence of $\underline{\underline{G}}(\underline{p})$ yields the Gauss-Newton optimization method

$$\underline{p}^{k+1} = \underline{p}^{k} - \left[\underline{\underline{J}}(\underline{p}^{k})^{\mathsf{T}} \underline{\underline{J}}(\underline{p}^{k})\right]^{-1} \underline{\underline{J}}(\underline{p}^{k})^{\mathsf{T}} \underline{r}(\underline{p}^{k}). \tag{4.13}$$

The Jacobian required for the Gauss-Newton optimization method is assembled by partial derivatives of the strain field with respect to the parameter vector $\underline{p}$. These entries are estimated by the constitutive equation $\boldsymbol{\sigma} = \mathbb{C}(\boldsymbol{x})[\boldsymbol{\varepsilon}]$ while assuming the stress field is independent of $\underline{p}$

$$\frac{\partial \boldsymbol{\varepsilon}}{\partial p_i} = -\mathbb{C}^{-1} \frac{\partial \mathbb{C}}{\partial p_i} \mathbb{C}^{-1} [\boldsymbol{\sigma}] = -\mathbb{C}^{-1} \frac{\partial \mathbb{C}}{\partial p_i} [\boldsymbol{\varepsilon}] . \tag{4.14}$$

This assumption was successfully used in the identifications of homogeneous parameters and showed significant increase performance (Lecompte et al., 2007). In the context of this work, the comparison to a finite difference scheme of $\partial\varepsilon/\partial p_i$ showed a difference in the converged material properties of up to 10%. Therefore, the solution with the analytic derivative (Equation (4.14)) was used as a starting value for the optimization with the finite difference scheme, which had a significant higher computational cost because in each iteration, solutions for small fluctuations of p_i had to be calculated.

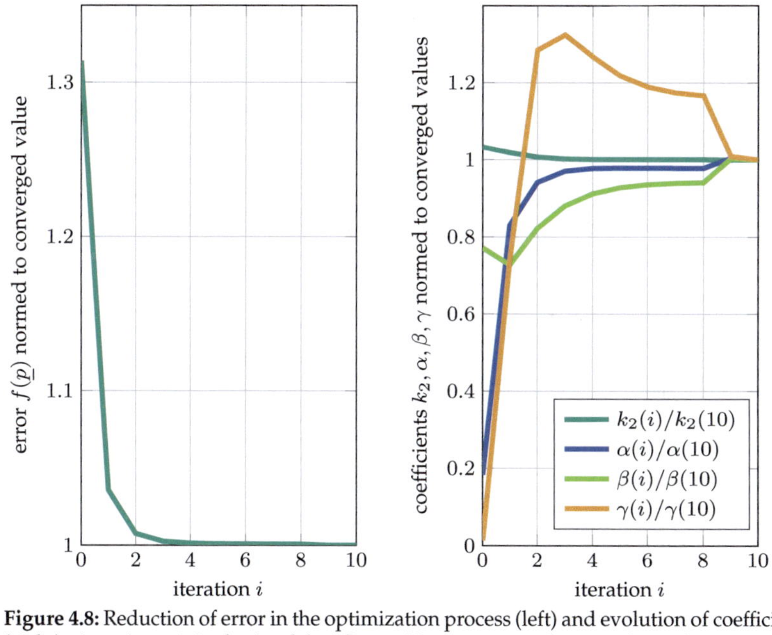

Figure 4.8: Reduction of error in the optimization process (left) and evolution of coefficients (right), iterations 1-8 obtained by Gauss-Newton procedure with analytic Jacobian (Equation (4.14)), iterations 9-10 obtained by finite difference scheme

As mentioned in above, due to the lack of reliable measurement techniques to measure the out of plane strain of the specimen, the parameter k_1 is not identifiable. Its value is obtained through a Mori-Tanaka (Mori and Tanaka, 1973) homogenization in the following way $k_1 = \frac{1}{3}\mathbb{P}_1 \cdot \mathbb{C}_{\mathrm{MT}}$.

Gradient based optimization algorithms are, generally, highly sensitive to a deviation of start values. Hartmann and Gilbert (2017) showed that for the case of homogeneous isotropic elasticity (surface strain and force measurement including some measuring uncertainties), the material parameters can be identified at sufficient precision. However, to minimize the risk of convergence problems, the start values to the Gauss-Newton method are obtained in the following way.

An inverse parameter identification is used to identify the homogeneous stiffness of the specimen. In the following step, the heterogeneous stiffness tensor (Section 4.3.1) is fitted to the homogeneous stiffness tensor using a least-square approach and a genetic optimization algorithm. The obtained values for $k_2, \alpha, \beta, \gamma$ are now used as starting values.

Figure 4.8 (left) shows the convergence behavior of the described optimization procedure. On average, the Gauss-Newton algorithm converges in less than 10 iterations, whereas convergence is achieved when the relative change of every parameter is below 1%. The plot in Figure 4.8 left shows the reduction of the error function $f(\underline{p})$ over the Gauss-Newton iterations i. An error reduction of about 25% was obtained. Figure 4.8 (right) shows the evolution of the four parameters k_2, α, β and γ normed to its converged values.

4.4 Results and discussion

Figure 4.9 and Figure 4.10 depict a comparison between measured and simulated strain fields for two different biaxial tensile specimens for the load case of equibiaxial tension. For Figure 4.9, the fiber volume fraction of the specimen is 22%, and for Figure 4.10 it is 29%. The left columns of both figures show the measured strain field by digital image correlation. The middle column shows the strain fields obtained by finite element simulations with the consideration of a heterogeneous material behavior, see Section 4.3.1 and 4.3.2. The right column shows finite element results for a homogeneous stiffness. This stiffness was identified in a state-of-the-art standard procedure (Schemmann et al., 2015b), that is similar to the heterogeneous one mentioned above.

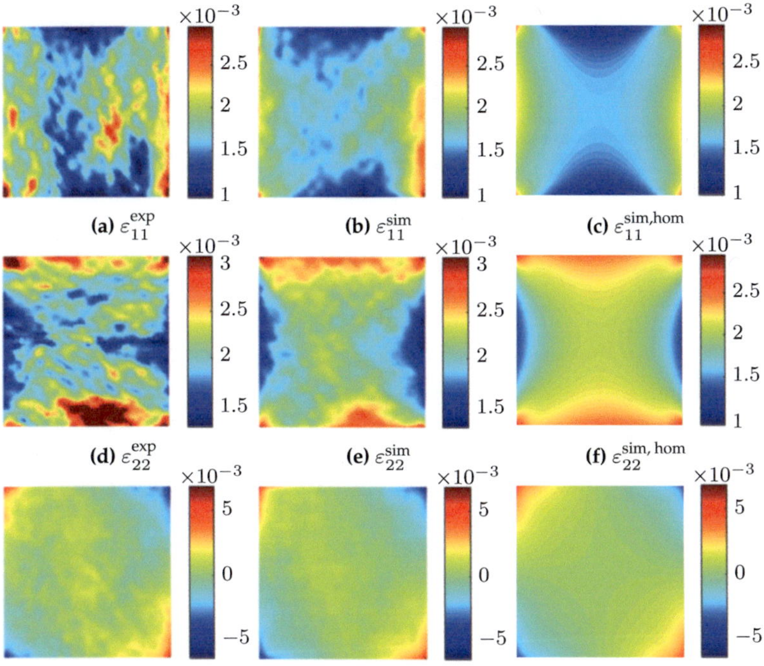

(a) $\varepsilon_{11}^{\text{exp}}$ **(b)** $\varepsilon_{11}^{\text{sim}}$ **(c)** $\varepsilon_{11}^{\text{sim,hom}}$

(d) $\varepsilon_{22}^{\text{exp}}$ **(e)** $\varepsilon_{22}^{\text{sim}}$ **(f)** $\varepsilon_{22}^{\text{sim, hom}}$

(g) $\varepsilon_{12}^{\text{exp}}$ **(h)** $\varepsilon_{12}^{\text{sim}}$ **(i)** $\varepsilon_{12}^{\text{sim, hom}}$

Figure 4.9: Comparisons of the strain fields for the load case of equibiaxial tension. The fiber volume fraction of the specimen is 22%. Left column: measured strain field by digital image correlation. Middle column: fitted under the assumption of linearity between the fiber orientation tensor and stiffness tensor. Right: fitted with a homogeneous stiffness over the specimen

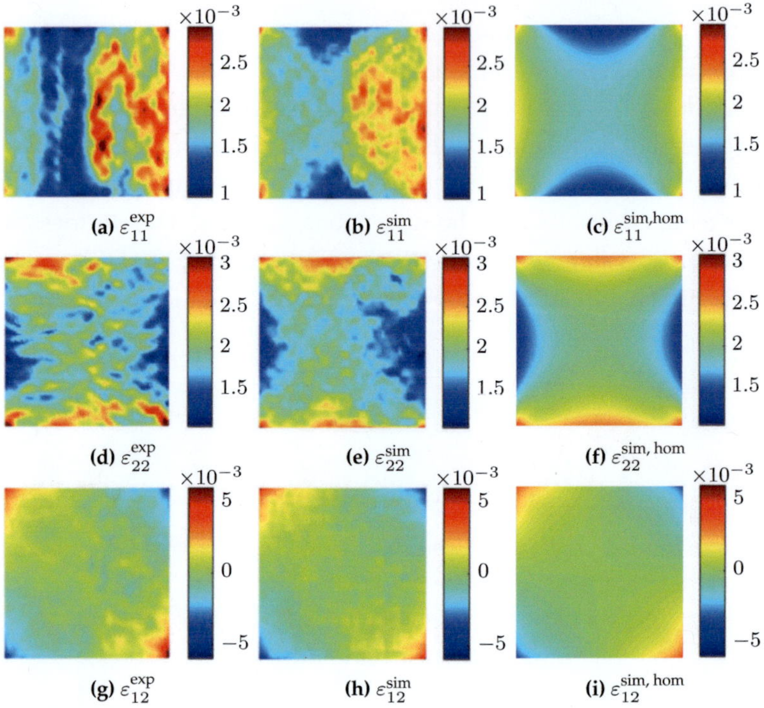

Figure 4.10: Comparisons of the strain fields for the load case of equibiaxial tension. The fiber volume fraction of the specimen is 29%. Left column: measured strain field by digital image correlation. Middle column: fitted under the assumption of linearity between the fiber orientation tensor and stiffness tensor. Right: fitted with a homogeneous stiffness over the specimen

Measurement errors at low strains such as 0.2-0.5%, and the evaluation of only five specimens (two are only shown within this chapter) requires to exercise caution to avoid over-interpretation of the results. Some to the authors reliable findings are, however, collected in the following.

First, the measured and homogeneously simulated strain fields show significant fluctuations which additionally motivates an identification of inhomogeneous material properties.

The new approach (middle column) covers tendencies of these fluctuations. The normal strains in Figure 4.9 and 4.10 seem to reproduce the measured strain field not only qualitatively but also quantitatively. A slice of the μCT scan of the specimen in Figure 4.10 is depicted in Figure 4.1. The μCT scan shows a typical SMC fiber orientation on the left two third of the specimen, whereas on the right side of the specimen, the mold-flow induced an increased amount of fiber bending. The consideration of the measured strains in the e_1-direction shows an increase on the right side. This suggests a lower stiffness in this area which may derive from the microstructure described above. Similarly, on the lower left side of the image, the fibers seem to be orientated dominantly horizontal, therefore reduction of the stiffness in the vertical direction could explain the higher strains.

4.5 Conclusions

In contrast to common testing procedures, we introduced a method to characterize the microstructure based heterogeneity of SMC in biaxial tensile tests. The micromechanically-motivated assumption of linearity between stiffness and fiber orientation distribution reduced the problem dimension to four identifiable parameters, even for a highly heterogeneous fiber orientation distribution. Volumetric compression is not measurable. However, all other components of the stiffness tensor were obtained in a single biaxial tensile specimen (plane stress) in one single experiment.

Compared to traditional homogenization schemes, the introduced approach requires one measurement per fiber volume fraction and aspect ratio to predict the stiffness for any fiber orientation distribution. New measurements are required for a different fiber volume fraction or fiber aspect ratio. Homogenization approaches on the other hand do not require tests of the composite, however the elastic properties of the phases are needed as an input.

Chapter 5

Cruciform specimen design for biaxial tensile testing of SMC[1]

5.1 Introduction

5.1.1 Motivation

With their high mass-specific strength and stiffness, fiber reinforced polymers receive considerable attention in mass-reduction strategies in the automotive and aerospace sector. The stiffest and strongest class of composites are continuous fiber reinforced composites. This work, however, focuses on discontinuous fiber reinforced composites in the form of of sheet molding compounds (SMC). SMC offer a large freedom of design and the low cycle times allow for mass production. Characteristics of SMC include brittleness, process-induced inhomogeneity and anisotropy, and comparably large microstructure dimensions.

The application of SMC is hindered by the lack of a precise understanding of its mechanical behavior. The focus of the present work lies on biaxial tensile testing of SMC which is driven by two main factors. Firstly, the characterization under biaxial stress states covers

[1] This chapter is based on the paper "Cruciform specimen design for biaxial tensile testing of SMC" (Schemmann et al., 2018c).

a wide range of application loads on typical shell-like SMC structures. Secondly, as the implementation of virtual process chains and, especially, damage modeling increase in popularity, a detailed characterization of the damage behavior is essential to understand the complex damage phenomena and calibrate corresponding constitutive models. For instance, biaxial tensile testing allows for a validation and parameter identification of anisotropic damage models.

The cruciform specimen design significantly influences the characterization range of stress states as well as the precision of the characterized material properties. Optimization of specimen design is a multi-objective task that depends on the considered material, e.g., by means of mechanical behavior such as ductility or anisotropy, manufacturing restrictions, and microstructure length scales.

5.1.2 Chapter structure

This work presents an investigation of different specimen designs for biaxial tensile testing of SMC. We aim for an optimal geometry for the characterization of damage evolution under a wide range of biaxial planar stress states. The outline is as follows: Section 5.2.2 presents the background of biaxial tensile testing and introduces the two loading scenarios we apply for the experimental evaluation of the specimen design. Section 5.3 begins with a definition of our specific specimen optimality criteria. In the following, we introduce four specimen designs and evaluate them experimentally and partly with finite element simulations. Section 5.4 discusses and compares the results in terms of the introduced optimality criteria.

5.2 Materials and experiment

5.2.1 Materials and manufacturing process

The SMC considered here consists, specifically, of an unsaturated polyester polyurethane hybrid (UPPH) matrix system, reinforced with $23\text{vol.}\%$ glass fibers. The SMC was manufactured at the Fraunhofer Institute of Chemical Technology (ICT) at Pfinztal, Germany. The detailed composition of the matrix system is listed in Table 5.2. The length of the fibers is 25.4mm.

component	trade name	weight fraction
UPPH resin	Daron ZW 14142	77%
adherent and flow aids	BYK 9085	1.5%
impregnation aid	BYK 9076	2.3%
deaeration aid	BYK A-530	0.38%
inhibitor	pBQ	0.0023%
peroxide	Trigonox 117	0.77%
isocyanate	Lupranat M20R	18%

Table 5.1: Composition of the UPPH resin (Hohberg et al., 2017a)

The SMC pre-impregnated fibers were manufactured on a belt conveyor system. The initial charge had 60% mold coverage (with exception of the specimens presented in Section 5.3.4) in a square mold with the dimensions $455 \times 455\text{mm}$ and consisted of three layers of pre-impregnated fibers. With regard to more details of the manufacturing process, we refer to Hohberg et al. (2017a). All specimens were cut with a water jet cutter from the center region of the plate to minimize the influence of inhomogeneous fiber dispersion, e.g. investigated by Asadi et al. (2016).

5.2.2 Biaxial tensile experiments

Fundamentals of biaxial tensile testing

In this section, we briefly introduce our biaxial tensile testing device at the Institute of Engineering Mechanics (KIT) and some fundamentals on biaxial tensile testing. The biaxial tensile testing device shown in Figure 5.1a consists of four horizontally positioned elctro-mechanical actuators that are arranged perpendicular to each other. Each axis allows for a maximum load of 150kN and is equipped with a load cell. The deformation of the specimen is measured by an integrated optical strain measurement system, via the displacement of five points on the bottom of the specimen. The movement of these points is the input for the measurement of the strain load and midpoint control. The active midpoint control allows for bending-free load application, even in the case of heterogeneous specimens. On the upper side of the specimen, a speckle pattern allows for full-field strain measurements via digital image correlation. Due to the, in general, inhomogeneous stress and strain fields in the cruciform specimens the parameter identification is typically not as straight forward as, e.g., in uniaxial tensile tests. More details concerning inverse parameter identification are, e.g., covered in Mahnken and Stein (1996a;b); Cooreman et al. (2007); Lecompte et al. (2007); Schemmann et al. (2015b; 2018a).

Figure 5.1b depicts an exemplary cruciform specimen. If we assume the specimen as homogeneous, the opposing forces on the specimen are equal and given by F_1 in positive and negative e_1-direction and F_2 in positive and negative e_2-direction. The ratio of these two forces is known as the loading ratio

$$\Gamma = \frac{F_2}{F_1}. \tag{5.1}$$

The loading ratio of $\Gamma = 0$ defines uniaxial tension or compression in the $\boldsymbol{e}_1$-direction, whereas the loading ratio $\Gamma = 1$ defines equi-biaxial tension or compression.

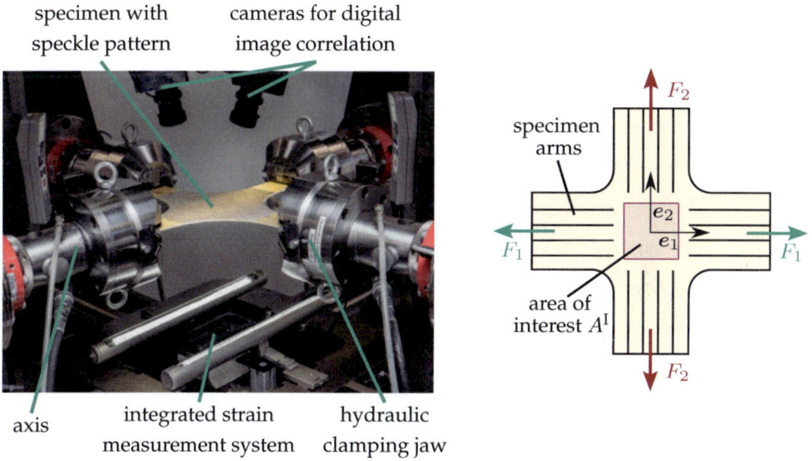

(a) Biaxial tensile testing device

(b) Cruciform specimen

Figure 5.1: Biaxial tensile testing setup and an exemplary cruciform specimen

We define the area of interest A^{I} as the measuring region which later can serve as input for the parameter identification. In all following contour plots, the area of interest is highlighted by a purple frame. For the cruciform specimen design, the area-averaged strain over the area of interest

$$\bar{\varepsilon}^{\mathrm{I}}_{ij} = \frac{1}{A^{\mathrm{I}}} \int_{A^{\mathrm{I}}} \varepsilon_{ij} \left(\boldsymbol{x}\right) \mathrm{d}A, \tag{5.2}$$

is a key quantity in the evaluation of the specimen geometries, with $\varepsilon_{ij} = \left(\partial u_i / \partial x_j + \partial u_j / \partial x_i\right) / 2$ being the components of the infinitesimal strain tensor defined from the displacement vector u_i.

Experimental procedures

For the experimental and simulative evaluation of the specimen designs, we focus on two loading scenarios. We avoid compressive loads as they may lead to buckling issues. The distinction of the mechanical phenomena like, for instance, elasticity, stiffness degradation (damage), plasticity, and viscoelasticity is simplified by a stepwise load application and waiting times between the load application steps.

The first procedure $P_{\Gamma=0}$ is based on cyclic uniaxial tension in e_1-direction with a stepwise load increase. Between these loading steps, we apply a constant uniaxial tension in the e_2-direction. This allows for an observation of stiffness degradation in and perpendicular to the main loading direction. Figure 5.2 shows a schematic stress and strain path for the loading procedure $P_{\Gamma=0}$. The tensile load is applied strain-controlled. Perpendicular to the respective loading direction, a force control ensures a low contact force.

The second procedure $P_{\Gamma=1}$ (see Figure 5.3) is designed such that the damage inducing load is equi-biaxial tension. Between the biaxial tension steps of increasing amplitude, uniaxial tension steps of small and constant amplitude are applied in the e_1-direction and subsequently in the e_2-direction in order to estimate the stiffness evolution in these directions. The strain and force control is analogous to the first procedure.

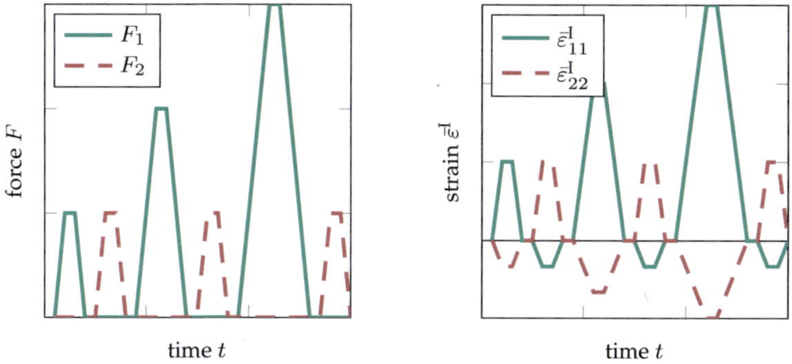

(a) Force path for procedure $P_{\Gamma=0}$ **(b)** Strain path for procedure $P_{\Gamma=0}$

Figure 5.2: Schematic force and strain paths for experimental procedure $P_{\Gamma=0}$

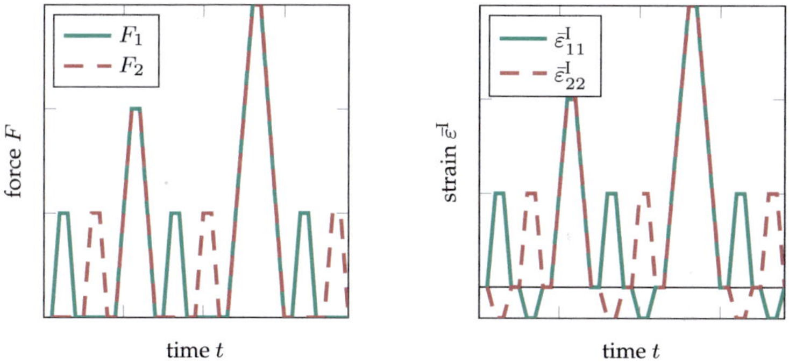

(a) Force path for procedure $P_{\Gamma=1}$ **(b)** Strain path for procedure $P_{\Gamma=1}$

Figure 5.3: Schematic force and strain paths for experimental procedure $P_{\Gamma=1}$

We chose these two procedures, because we assume that the specimens which perform well – in the sense of allowing to detect an anisotropic damage evolution – at the extreme loading ratios $\Gamma = 0$ and $\Gamma = 1$, also perform well at the loading ratios in between.

5.3 Specimen designs and experimental results

5.3.1 Specimen requirements

Our goal for biaxial tensile tests is the robust measurement of the anisotropic evolution of damage and material strength of SMC under various planar stress states. Smits et al. (2006) listed criteria for optimal cruciform specimens of unidirectional composites. We modified and extended this list by taking the specific properties of SMC into consideration. We believe, that the optimality of a specimen is qualitatively determined by the following criteria:

1. *Wide range of achievable stress states.* An optimal specimen geometry allows for all biaxial tension stress states, i.e. ratios between normal stresses. By coordinate transformation then all planar stress states (not in magnitude, but in relation to each other) are covered. The measurement of all stress states with the same geometry does not only ensure a good comparability in contrast to multiple specimen geometries, but also allows for a straight forward application of non-monotonic loading paths.
2. *Damage dominantly in the area of interest.* Since it is our goal to inspect damage in the area of interest, we would like to avoid premature specimen failure in the arms, and thus analyze the material behavior at highest possible strains in the area of interest.
3. *Homogeneity of stress state in the area of interest.* For the analysis of damage, it is desirable to reach a homogeneous stress state in the area of interest. This implies the demand to avoid stress concentrations.
4. *Robust parameter identification.* The parameter identification must be a well-posed problem and robust with respect to noise of measured quantities (forces and strain field) (Hartmann and Gilbert, 2017). A robust parameter identification is essential for reproducibility.

The robustness of the parameter identification is, however, not considered in this thesis.

5. *Large area of interest.* The microstructural dimensions are in case of SMC, compared to other discontinuous fiber reinforced polymers, relatively large. The typical fiber roving length is $25.4\mathrm{mm}$, whereas one roving is assembled of thousands of filaments. As the specimen size is limited, it is our goal to achieve a considerably large area of interest.
6. *Low production effort.* In contrast to uniaxial tensile specimens, the load ratio is an additional parameter to be considered in the design of experiments. SMC is known to show significant scatter in experimental results. Additionally, the anisotropy and inhomogeneity must be considered in the design of experiments. The resulting high number of required experiments can better be coped with, if the economical effort for the specimen production is low.

5.3.2 Unreinforced specimen arms

Specimen design

We introduce the first specimen design in line with the ISO 16842 (2014) norm for biaxial tensile testing of sheet metals. Typical metals analyzed in biaxial tensile testing are aluminum and deep-drawing steels. These materials show, in contrast to SMC, large strain in the nonlinear (plastic) regime as well as work hardening. Figure 5.4 shows the corresponding design of a subsequently tested SMC specimen. The main advantages of water jet cutting include excellent heat removal and a minimum slit width of $1.2\mathrm{mm}$. Aiming at a maximum surface area in the arms, we decided to introduce only 3 slits which lead to a reduction of the arm surface area of 7.2%. In contrast to the normed design, we chose the slits to extend along the entire arms, in order to avoid a damaged area enforced by the starting points of water jet cutting.

Figure 5.4: Image of the unreinforced cruciform specimen, thickness of the area of interest: 2mm

Results

For both, the testing procedure $P_{\Gamma=0}$ and the testing procedure $P_{\Gamma=1}$, the specimens failed in the arms before a sufficiently high strain in the area of interest could be reached. Figure 5.5a shows the force over the strain in the e_1-direction for testing procedure $P_{\Gamma=0}$. The corresponding contour plot (see Figure 5.5b) shows that the strain in the arms is approximately equal to the strain in the area of interest, which violates optimality criterion 2. Figure 5.6a shows the mean values of the forces and the strains in the e_1-direction and the e_2-direction for testing procedure $P_{\Gamma=1}$. The contour plot (Figure 5.6b) shows that the normal strain ε_{11} is significantly higher in the arms than in the area of interest. The plot only shows the areas of positive strains. The asymmetry of the image section is due to the recording area of the cameras. A contour plot of the area of interest at the instant shortly before failure and an image of the failed specimen for all specimen designs can be found in Appendix B.

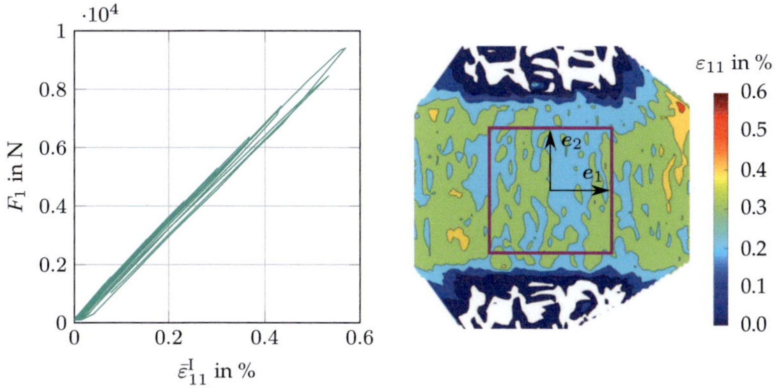

(a) Force F_1 in $\boldsymbol{e}_1$-direction over averaged normal strain $\bar{\varepsilon}^{\mathrm{I}}_{11}$ in the area of interest

(b) Contour plot of the normal strain ε_{11} at $\bar{\varepsilon}^{\mathrm{I}}_{11} = 0.3\%$

Figure 5.5: Unreinforced specimen arms with loading scenario $P_{\Gamma=0}$

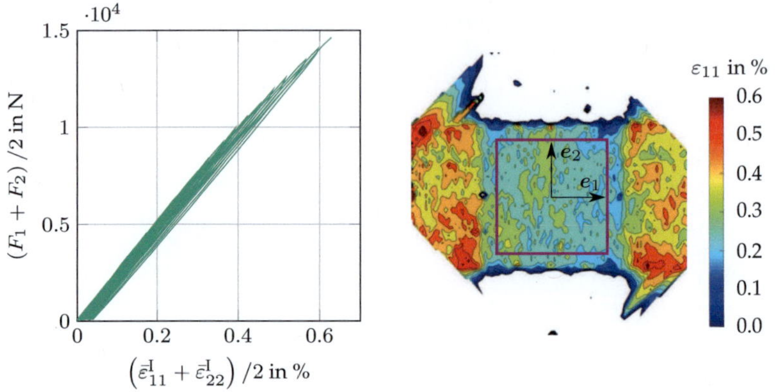

(a) Force over strain, average values over the e_1- and the e_2-direction

(b) Contour plot of the normal strain ε_{11} at $\bar{\varepsilon}^{\mathrm{I}}_{11} = 0.3\%$

Figure 5.6: Unreinforced specimen arms with loading scenario $P_{\Gamma=1}$

5.3.3 Bonded reinforcements on the arms

Specimen design

To avoid premature failure in the arms, we reinforced the specimen by bonding strips on the arms as shown in Figure 5.7. The upper part of the picture shows the fixation of the reinforcing strips during the manufacturing process. The strips are cut from the same SMC plate as the specimen. We glued with epoxy adhesive which has similar material properties as the resin material (UPPH). The preparation of the specimens requires many hours of manual work and, therefore, violates optimality criterion 6.

Figure 5.7: Image of the cruciform specimen with bonded reinforcements on the arms, thickness of the area of interest: 2mm

Results

The bonded reinforcements significantly increase the maximum achievable strain. Here, the specimens also failed in the arms, however, a higher strain in A^{I} can be reached compared to the unreinforced specimen. One reason for the premature failure in the arms was a partial detachment of the bonded reinforcing strips. The force strain diagram (see Figure 5.8a) demonstrates the increase of the maximum reachable strain for testing procedure $P_{\Gamma=0}$. Besides that, a stiffness degradation is already visible from the decrease of the slope of the force strain curve after each cycle. However, the contour plot (5.8b) shows that the strain in the arms is still approximately equal to the strain in the area of interest. Figure 5.9a shows the mean values of the forces and the strains in the e_1-direction and the e_2-direction for the testing procedure $P_{\Gamma=1}$. Here too, the maximum strain is slightly higher compared to the unreinforced specimen. The contour plot (Figure 5.9b) shows that the normal strain in the arms is still significantly higher than the normal strain in the area of interest.

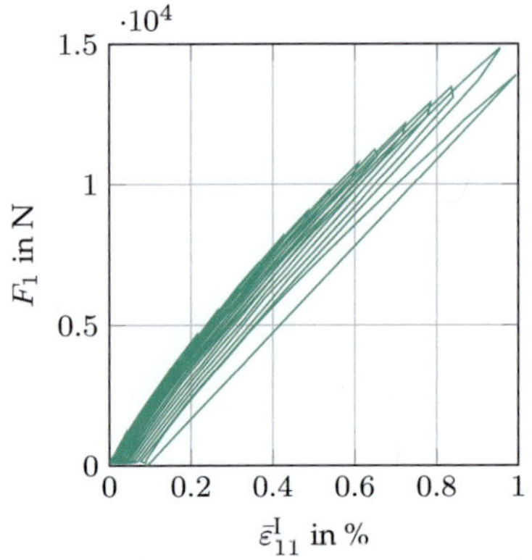

(a) Force F_1 in $\boldsymbol{e}_1$-direction over averaged normal strain in the area of interest $\bar{\varepsilon}^{\mathrm{I}}_{11}$

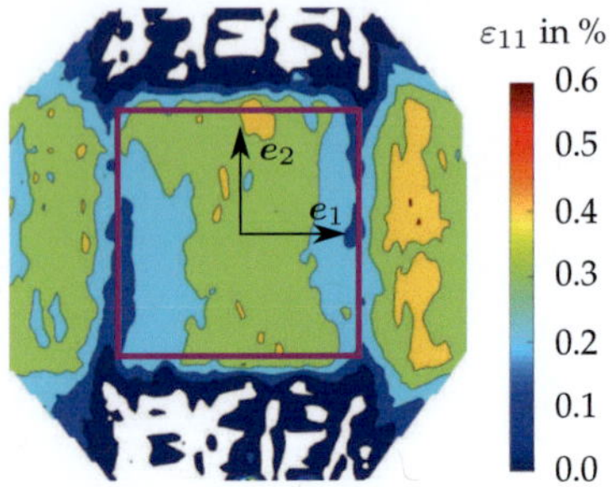

(b) Contour plot of the normal strain ε_{11} at $\bar{\varepsilon}^{\mathrm{I}}_{11} = 0.3\%$

Figure 5.8: Bonded reinforcements on the arms with loading scenario $P_{\Gamma=0}$

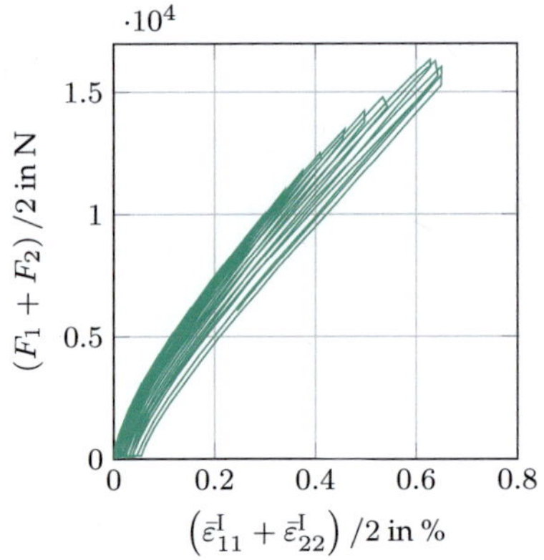

(a) Force over strain, average values over the $\boldsymbol{e}_1$- and the $\boldsymbol{e}_2$-direction

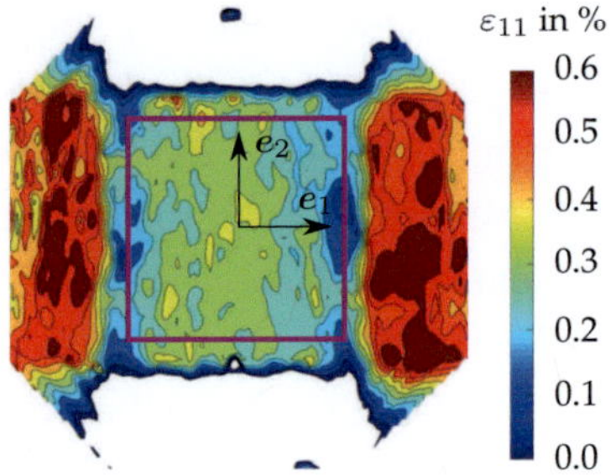

(b) Contour plot of the normal strain ε_{11} at $\bar{\varepsilon}^{\mathrm{I}}_{11} = 0.3\%$

Figure 5.9: Bonded reinforcements on the arms with loading scenario $P_{\Gamma=1}$

5.3.4 Continuous fiber reinforced arms

Particularities in specimen manufacturing

To overcome the disadvantages of bonded reinforcements on the arms, we considered the manufacturing process of SMC with unidirectional tapes (Bücheler et al., 2017). The unidirectional tapes' constituents are UPPH resin (also in SMC) and 60vol.% carbon fibers. The plates were manufactured with 100% mold coverage to avoid in-mold slip of the layers during the compression molding. More details regarding the co-molding process can be found in Bücheler (2017). Figure 5.10 shows the symmetric layup, where the middle layer consists of SMC and the outer layers are tapes with fiber orientations perpendicular to each other.

Figure 5.10 and Figure 5.11 depict the pockets machined into the continuous layers of the specimen. The milling process of these pockets is described in the following.

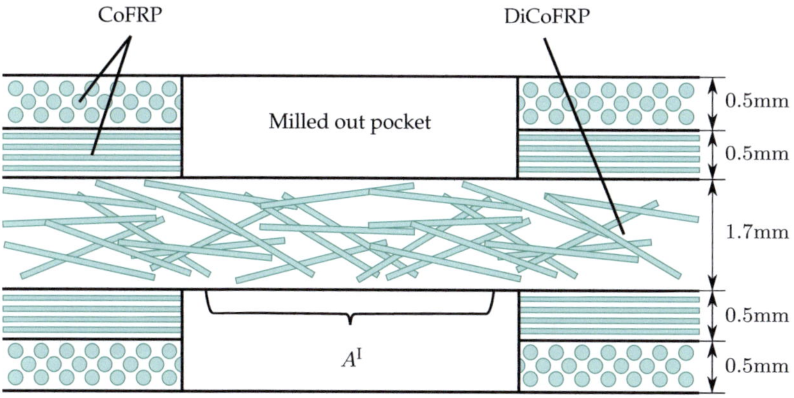

Figure 5.10: Cross section of continuous fiber reinforced SMC, with milled pockets in the area of interest

For this process, a diamond coated end mill with 8 cutting edges and a multi-cut geometry by Hufschmied GmbH (Bobingen, Germany) was used. The specimens were clamped into position with a line clamping on two sides of the pocket as to minimize the distance of the pocket center to either side of the clamping. This clamping strategy also allowed for a continuous process, i.e., no reclamping was necessary. We generated the tool path from a CAD model of the specimens with the Siemens NX manufacturing CAM system. The values of the cutting speed v_c and the feed per tooth f_{PT} were experimentally qualified during preliminary tests with a similar material and the same tool. These parameters were set at $v_c = 66\text{m/min}$ and $f_{PT} = 0.07\text{mm}$, respectively. The main challenges in machining of the pockets were to prevent any kind of delamination which is critical in plunging into the material, and to obtain a smooth, even surface without heat-induced damages or altered material properties. After the first cut with a depth of 1mm, in some areas of the pocket, remnants of the continuous fibers were found. These remnants likely originate from a movement of the unidirectional tapes in e_3-direction during the compression molding process. Nevertheless, further machining steps were performed in 0.1mm steps to ensure a pure SMC sector in the area of interest.

Specimen design

The novel manufacturing technique significantly increases the arms' stiffness and strength in the loading directions and, therefore, defines new constraints in the search for the optimal geometry. Here, we introduce two geometries which we consider to be a good compromise between the optimality criteria presented in Section 5.3.1. As there are difficulties in parametric shape optimization (Bauer et al., 2016), we decided to perform a high number (thousands) of FE simulations and thoroughly selected suitable solutions manually.

The first geometry, in the following called geometry 1, follows the design of the unreinforced specimen. Figure 5.11a shows an image of the design. The significant increase of arm strength allowed to incorporate six slits in each arm to reduce the peak stress at the slit's ends and reduce load transition in the slits perpendicular to the normal stress direction. The slits end slightly before the beginning of the milled out pockets.

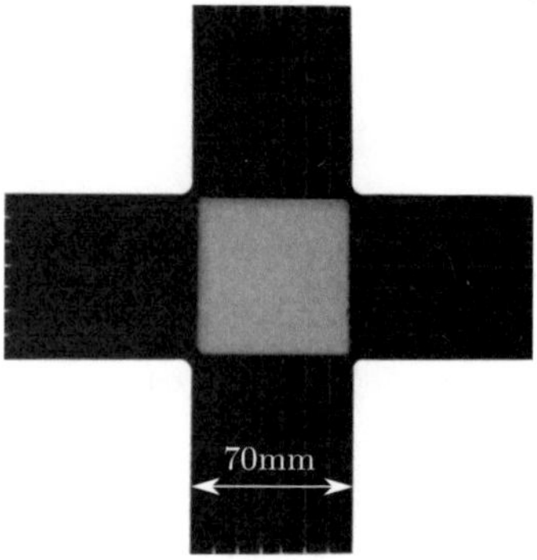

(a) Geometry 1

(b) Geometry 2

Figure 5.11: Images of specimens with continuous fiber reinforced arms, thickness of the area of interest: 1.7mm

Figure 5.12 depicts strain fields computed from finite element simulations on geometry 1. We took advantage of the specimen's symmetry and only simulated one fourth of the geometry with a fine discretization of shell elements. As we were only interested in trends of different geometry modifications and not precise strain fields, we assumed linear elastic isotropic material behavior with a different stiffness and element thickness in the pocket and the reinforced area. We would like to remark here, that a precise quantitative comparison of designs should include the prediction of macroscopic cracks leading to macroscopic failure, and an evaluation of the observability of the damage stage. This was not achievable within the scope of the presented research. The upper images show the strain distributions for a uniaxial loading scenario $P_{\Gamma=0}$ in the horizontal $\boldsymbol{e}_1$-direction. The strain distribution in the area of interest shows only small fluctuations. The outer arms contribute more to the load transition into the specimen, as part of the outer load is transmitted into the tape-reinforced regions. The lower images visualize the strain distribution for an equi-biaxial tensile loading scenario $P_{\Gamma=1}$. The distribution of the normal strain ε_{11} (Figure 5.12c) shows peaks at the end of the horizontal slits. Figure 5.12d depicts the distribution of the maximal principal strain ε_1. Due to the reduced stiffness in the area of interest, here, the strain is significantly higher than in the earlier discussed designs.

Geometry 2 is different from geometry 1 in such a way that the milled out area has curved edges (see Figure 5.11b). In analogy to uniaxial tapered bone specimens, our aim was to achieve an elevated stress level in the area of interest. Figure 5.13 shows the strain fields for uniaxial and biaxial load. For the uniaxial loading case, the stress distribution is more homogeneous, however, in front of the second outer slit end, there is a significant strain concentration. For the case of equi-biaxial loading, the strain concentrations, especially the principal strain ε_1 (see Figure 5.13d), shows significant peaks towards the end of the slits.

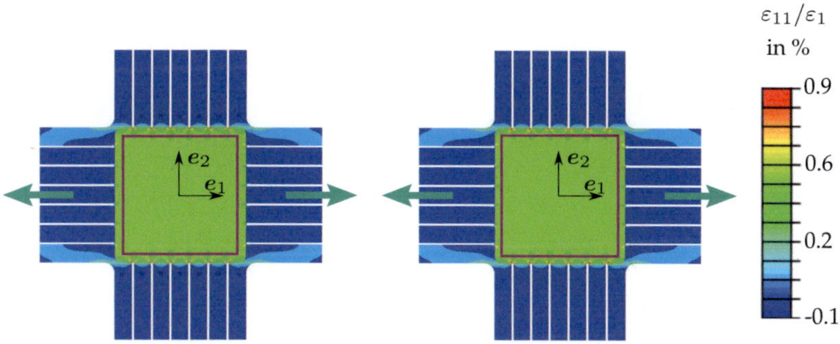

(a) ε_{11} for uniaxial load in $\boldsymbol{e}_1$-direction **(b)** ε_1 for uniaxial load in $\boldsymbol{e}_1$-direction

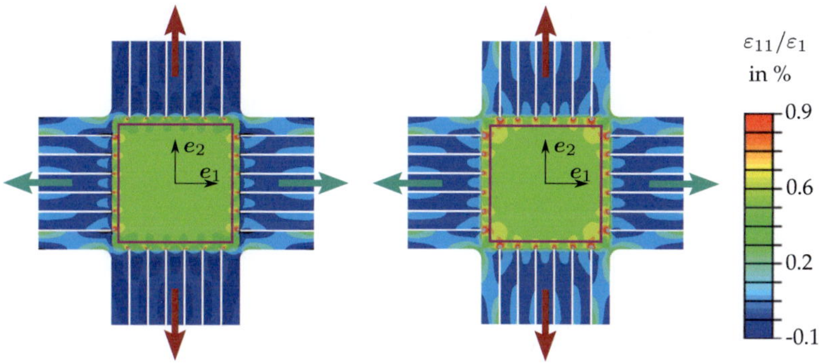

(c) ε_{11} for equi-biaxial load **(d)** ε_1 for equi-biaxial load

Figure 5.12: FE results for geometry 1 specimen at a macroscopic strain of $\bar{\varepsilon}^{\mathrm{I}}_{11} = 0.27\%$. The strain fields show the normal strain ε_{11} and the largest principal strain ε_1.

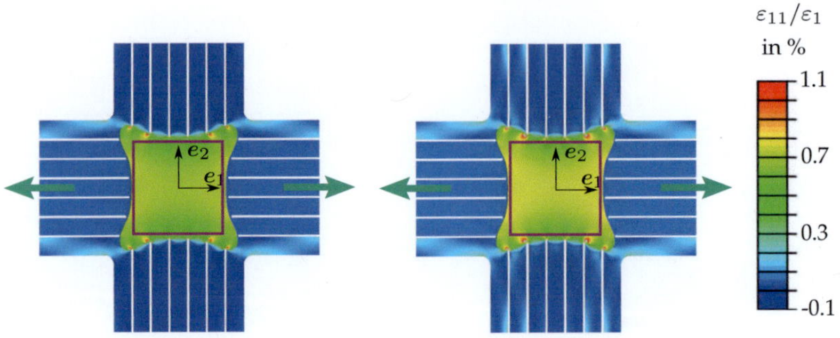

(a) ε_{11} for uniaxial load in $\boldsymbol{e}_1$-direction **(b)** ε_1 for uniaxial load in $\boldsymbol{e}_1$-direction

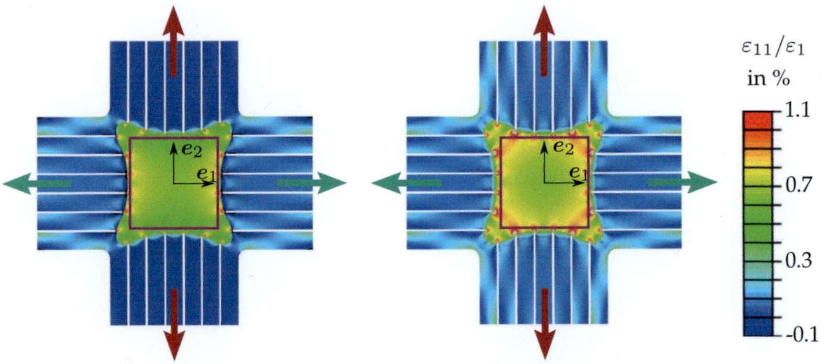

(c) ε_{11} for equi-biaxial load **(d)** ε_1 for equi-biaxial load

Figure 5.13: FE results for geometry 2 specimen at a macroscopic strain of $\bar{\varepsilon}^{\mathrm{I}}_{11} = 0.27\%$. The strain fields show the normal strain ε_{11} and the largest principal strain ε_1

Results

The contour plots in Figure 5.14 through Figure 5.17 show that the experimental strain distributions are similar to those predicted from the simulations. Figure 5.14b confirms the relatively homogeneous strain field of Figure 5.12a for geometry 1 and testing procedure $P_{\Gamma=0}$. For testing procedure $P_{\Gamma=1}$, the small strain peaks at the end of the slits mentioned in conjunction with Figure 5.12c are visible again (see Figure 5.15b). For both testing procedures, the strain in the area of interest is significantly higher than the strain in the arms. Failure occurs in the area of interest when a sufficiently high strain is reached. Figure 5.14a shows the force over the strain in the e_1-direction for the testing procedure $P_{\Gamma=0}$. This figure demonstrates that the continuous fiber reinforced layers significantly increase the maximum achievable strain. A stiffness degradation is clearly visible. Figure 5.15a shows that the reinforcement increases the maximum achievable strain also for testing procedure $P_{\Gamma=1}$.

Figure 5.16b and Figure 5.17b show the contour plots of geometry 2. Here too, the strain in the area of interest is significantly higher than in the arms. The strain peaks at the end of the slits, as mentioned for Figure 5.13, are visible, particularly pronounced for testing procedure $P_{\Gamma=1}$. Failure occcurs in the area of interest, but is initiated at the end of the slits. The force strain diagrams (Figures 5.16a and 5.17a) show that also geometry 2 increases the maximum achievable strain, but slightly less than geometry 1.

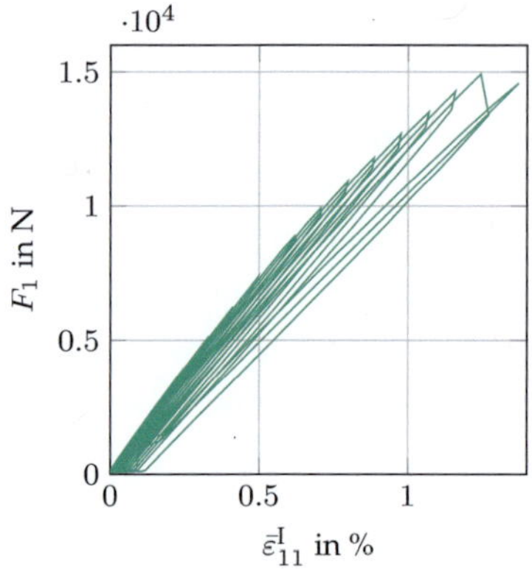

(a) Force F_1 in $\boldsymbol{e}_1$-direction over averaged normal strain $\bar{\varepsilon}^{\mathrm{I}}_{11}$ in the area of interest

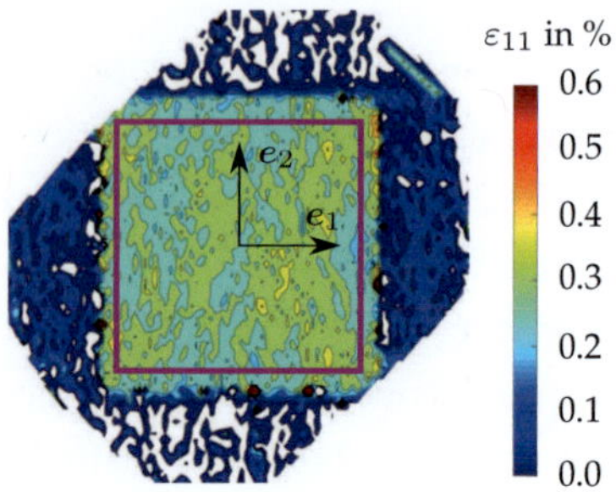

(b) Contour plot of the normal strain ε_{11} at $\bar{\varepsilon}^{\mathrm{I}}_{11} = 0.3\%$

Figure 5.14: Geometry 1 with continuous fiber reinforced arms with loading scenario $P_{\Gamma=0}$

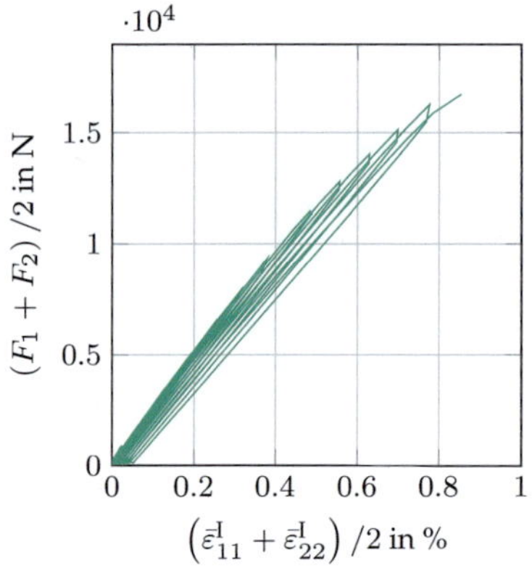

(a) Force over strain, average values over the $\boldsymbol{e}_1$- and the $\boldsymbol{e}_2$-direction

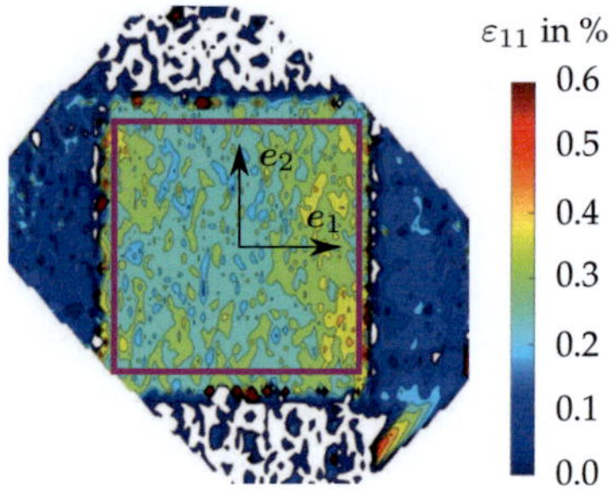

(b) Contour plot of the normal strain ε_{11} at $\bar{\varepsilon}^{\mathrm{I}}_{11} = 0.3\%$

Figure 5.15: Geometry 1 with continuous fiber reinforced arms with loading scenario $P_{\Gamma=1}$

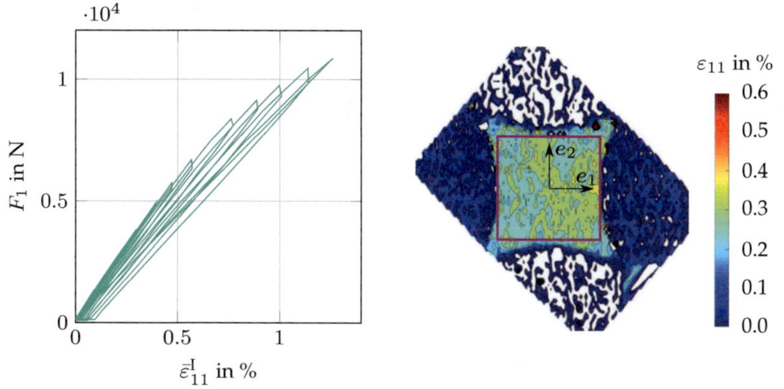

(a) Force F_1 in $\boldsymbol{e}_1$-direction over averaged normal strain $\bar{\varepsilon}^{\mathrm{I}}_{11}$ in the area of interest

(b) Contour plot of the normal strain ε_{11} at $\bar{\varepsilon}^{\mathrm{I}}_{11} = 0.3\%$

Figure 5.16: Geometry 2 with continuous fiber reinforced arms with loading scenario $P_{\Gamma=0}$

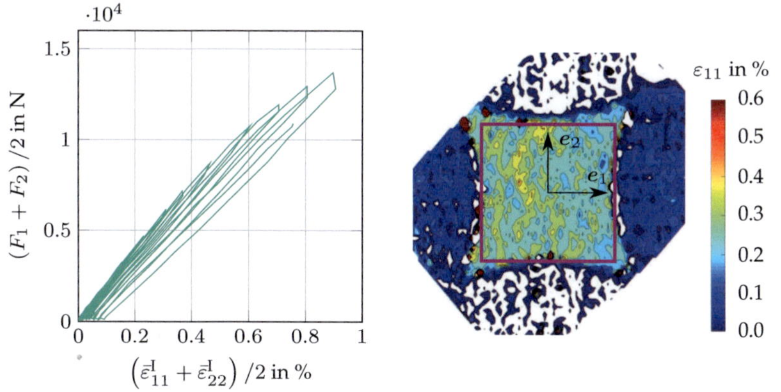

(a) Force over strain, average values over the $\boldsymbol{e}_1$- and the $\boldsymbol{e}_2$-direction

(b) Contour plot of the normal strain ε_{11} at $\varepsilon^{\mathrm{I}}_{11} = 0.3\%$

Figure 5.17: Geometry 2 with continuous fiber reinforced arms with loading scenario $P_{\Gamma=1}$

5.4 Discussion

Table 5.2 summarizes the maximum attained normal strains averaged over the area of interest $\max\left(\bar{\varepsilon}^{\mathrm{I}}_{11}\right)$ in loading scenario $P_{\Gamma=0}$ and $\max\left(\bar{\varepsilon}^{\mathrm{I}}_{11}+\bar{\varepsilon}^{\mathrm{I}}_{22}\right)/2$ in loading scenario $P_{\Gamma=1}$ accordingly. These strains especially quantify the level of achievable strain in the area of interest and thus allow to evaluate the specimen with regard to criterion 2, since damage only occurs if sufficiently high strains are achieved (criterion 2 of the previously defined specimen requirements (Section 5.3.1)). The ranking of the specimens is equal for both loading scenarios. The unreinforced specimen fails at the lowest strains, bonded reinforcements on the arms lead to a significant improvement, the continuous fiber reinforced specimens perform better, whereas geometry 1 performs slightly better than geometry 2. In loading scenario $P_{\Gamma=0}$, the continuous reinforced geometry 1 reaches 87% of the failure strain measured in a uniaxial tensile bone specimen tested on the same machine. In parallel to the maximum strain reached in the area of interest, a steady reduction of the slope in the force-strain plots indicates the evolution of damage. A visual comparison of these plots indicates a good agreement with Table 5.2, such that, the specimens which allow for the observation of higher strains also show a higher slope reduction in the force-strain plot. The interpretation of the results requires great caution, since only one experiment per specimen geometry and load case was performed.

reinforcement type		none	bonded SMC	cont. geom. 1	cont. geom. 2	uniax. bone
$\max\left(\bar{\varepsilon}^{\mathrm{I}}_{11}\right)$	$P_{\Gamma=0}$	0.57%	1.00%	1.37%	1.26%	1.57%
$\max\frac{\bar{\varepsilon}^{\mathrm{I}}_{11}+\bar{\varepsilon}^{\mathrm{I}}_{22}}{2}$	$P_{\Gamma=1}$	0.63%	0.65%	0.85%	0.76%	-

Table 5.2: Maximum attained strain within one experiment for the different biaxial specimen designs and, for reference, the failure strain of a uniaxial bone specimen. Here, the operator $\max\left(\cdot\right)$ denotes the maximum value in the course of the experiment

Geometry 1 also performs equal or superior to the other specimens considering optimality criteria 1 and 3. Force F_1 is almost ideally transferred via a homogeneous normal stress in the e_1-direction, force F_2 governs the normal stress in the e_2-direction accordingly. Therefore, by a variation of the loading ratio Γ, we can achieve any planar tensile stress ratio. The strain state in the area of interest is rather homogeneous because of the high number of slits, that reduce the strain, as shown by DIC and FE simulations. Consequently, we may assume that (due to the slits) the disturbing effect of the arms on the stress distribution is weak and the stress field is likewise approximately homogeneous so that the stress state can be estimated from the forces divided by the material cross section in the area of interest. The thus obtained material parameters provide an estimate of the material behavior and can serve as a starting value for an inverse parameter identification (partly criterion 4).

When comparing geometry 1 and geometry 2 of the continuous fiber reinforced specimen, geometry 1 also shows advantages concerning the larger size of the area of interest (criterion 5). The manufacturing effort (criterion 6) is the lowest in the unreinforced specimen. The bonded specimen requires a large amount of manual work. The continuous fiber reinforced specimen preparation requires additional manufacturing techniques (co-molding and milling), but is more time-efficient compared to the bonded specimen.

One experiment per design and load scenario is, of course, not statistically representative. We admit that our experimental investigation with only one specimen may not provide a precise quantitative design estimate, however, the experiments are (i) plausible within each experiment and in comparison of all experiments, (ii) show a good agreement with the FE simulations (continuous fiber reinforced specimen). The experiments in this work, hence, provide a preliminary estimation of the specimens' suitability.

5.5 Conclusions

This work provided experimental as well as numerical results to support the selection of a suitable specimen design for the characterization of the damage behavior of SMC. The authors, however, did not aim to characterize the anisotropic damage behavior of SMC, as the number of experiments was significantly too low and the differentiation of damage, viscoelasticity, and plasticity in discontinuous fiber reinforced polymers (Brylka et al., 2018) along with their parameter identification are still not fully understood. One result could be an understanding of the anisotropic stress-strain behavior of SMC. Geometry 1 of the continuous fiber reinforced specimen performed superior for our application, i.e. the optimality criteria and their prioritization we aimed for. The main advantages are the highest achievable strain in the area of interest among all specimen, the homogeneity of the strain in the area of interest, the large area of interest, and a moderate manufacturing effort.

Chapter 6

Anisotropic mean-field modeling of debonding and matrix damage in SMC[1]

6.1 Introduction

6.1.1 Motivation and materials

With their high mass-specific strength and stiffness, fiber-reinforced polymers receive considerable attention in mass-reduction strategies in the automotive and aerospace sector. The strongest and stiffest composites are continuous fiber-reinforced composites. Discontinuous fiber-reinforced polymers, however, are more amenable to high-volume automotive applications because of their geometric freedom and significantly reduced cycle times in manufacturing processes. In this chapter, we focus on a specific discontinuous fiber-reinforced polymer: sheet molding compound (SMC). SMC consist of a thermoset matrix – in our case, epoxy or unsaturated polyester polyurethane hybrid (UPPH) resin – reinforced with long ($\approx$ 25mm) glass fibers. Parts are typically manufactured by compression molding of pre-impregnated

---

[1] This chapter is based on the paper "Anisotropic mean-field modeling of debonding and matrix damage in SMC composites" (Schemmann et al., 2018b).

fibers, also known as prepregs. The orientation evolution of fibers during the compression molding manufacturing process leads to a mold-flow-induced microstructure.

The demand for a precise and efficient prediction of the resulting inhomogeneous, anisotropic, process-dependent mechanical properties arises from industry's interest in shortening development cycles and reducing extensive and costly prototyping. A wide range of research is, therefore, associated to the virtual process chain of discontinuous fiber-reinforced composites, e.g., moldfilling and fiber orientation evolution analysis, prediction of warpage and eigenstresses, and structural analysis including damage.

6.1.2 Chapter structure

Despite the enormous work already conducted within the research of discontinuous fiber reinforced polymers, some deficiencies still remain open. Some of those deficiencies are addressed by the actually presented model. Few models, e.g., are physically motivated and take the microscale into account, but are still efficiently applicable to calculations of structural components (e.g., parts that are of interest to the industry (Bruderick et al., 2013; Jansen, 2013; Hangs et al., 2016). Hereby, a numerical regularization based on the redistribution of load from fibers with debonded interfaces to undamaged fibers can be helpful. To the authors' knowledge, a Weibull weakest link approach with a heterogeneous stress distribution on each interface was not presented so far. Added value lies in the simulation of non-proportional loading paths and a rigorous visualization of the resulting anisotropic evolution of the effective and the microstructural quantities (see Section 6.5.3). This allows for a better discussion and a better understanding of, e.g., damaged-induced anisotropy within SMC composites.

This chapter presents an anisotropic continuum mechanical mean-field damage model taking into account an arbitrary orientation distribution of straight fibers. The outline of this chapter is as follows: In Section 6.2, a mea-field Mori-Tanaka homogenization scheme based on an empirical fiber orientation distribution is presented. The model accounts for fiber-matrix interface debonding and matrix damage in SMC composites. In the latter, a maximum principal stress criterion is applied, whereas the former is based on a direction-dependent equivalent interface stress which is related to a Weibull survival probability. Hence, interface debonding is modeled by an anisotropic evolution of load-carrying fibers. Fiber breakage is neglected. Section 6.3 introduces the discretization procedure and numerical implementation of the damage models. A regularization approach ensures numerical and computational efficiency of the model. Section 6.4 deals with the identification of the parameters needed to properly describe the material model. Section 6.5 is devoted to the experimental validation.

6.2 Continuum mechanical model

6.2.1 SMC microstructure

Here, the SMC is treated as a two-phase composite consisting of a thermoset matrix phase Ω_{M} and glass fibers Ω_{F}. The matrix is characterized linear elastically by an isotropic matrix stiffness tensor $\mathbb{C}_{\mathrm{M}}$ and the corresponding volume fraction c_{M}. All fibers are modeled linear elastic with an isotropic stiffness $\mathbb{C}_{\mathrm{F}}$. Due to the low shear rates in the manufacturing process, fiber curvature and breakage during the manufacturing process are neglected. As shown, e.g., by Jendli et al. (2005), fiber breakage is the least important damage phenomenon in SMC composites under tension. In-situ experiments showed that fibers usually break only due to macroscopic crack propagation within the specimen. Additionally, Figure 6.2 shows, that for the material considered here, the fiber stress is beneath the 90% confidence intervall of the fiber strength. The fibers are, therefore, modeled as straight ellipsoids with a uniform aspect ratio a_{v}. The unit vector $\boldsymbol{n}$ describes the orientation of a fiber. The fiber volume fraction is $c_{\mathrm{F}} = 1 - c_{\mathrm{M}}$.

Figure 6.1 depicts a sliced micro-computed tomography (μCT) scan of SMC. The colors indicate the fiber orientation in each voxel following the orientation estimate of Pinter et al. (2018).

Figure 6.1: μCT scan of an SMC microstructure (UPPH resin, $c_{\mathrm{F}} = 23\%$, (Pinter, 2017)). Colors indicate fiber orientation in the corresponding voxel. Specimen dimensions are $10\mathrm{mm} \times 10\mathrm{mm}$.

The fiber filaments for the SMC manufacturing process are utilized in bundles of thousands of filaments. During the impregnation and compression molding, the fiber filaments partly disperse. Motaghi and Hrymak (2017) characterized the tow distortion in SMC. The remaining short-range order of fiber orientation and volume fraction is neglected since only the one-point statistic of the microstructure is considered.

The fiber orientation distribution function $f(\boldsymbol{n})$ (FODF) specifies the volume fraction $\mathrm{d}v/v$ of fibers with orientation $\boldsymbol{n}$ relative to the total fiber volume (see, e.g., Zheng and Du (2001); Müller and Böhlke (2016))

$$\frac{\mathrm{d}v}{v}(\boldsymbol{n}) = f(\boldsymbol{n})\,\mathrm{d}S. \tag{6.1}$$

Here, $\mathrm{d}S$ denotes the surface element on the unit sphere $S := \{\boldsymbol{n} \in \mathbb{R}^3 : \|\boldsymbol{n}\| = 1\}$. The FODF is non-negative, normalized, and symmetric

$$f(\boldsymbol{n}) \geq 0, \qquad \int_S f(\boldsymbol{n})\mathrm{d}S = 1, \qquad f(\boldsymbol{n}) = f(-\boldsymbol{n}), \qquad \forall \boldsymbol{n} \in S. \tag{6.2}$$

The FODF represents a one-point correlation function of the microstructure and is, therefore, the most simple statistical description of the fiber dominated microstructure. A distributional representation of $f(\boldsymbol{n})$ with K vectors $\boldsymbol{n}_\beta$ and corresponding weights $c(\boldsymbol{n}_\beta)$ is given by an empirical definition of the fiber orientation distribution with the Dirac delta function $\delta(\boldsymbol{n}, \boldsymbol{n}_\beta)$

$$f(\boldsymbol{n}) = \sum_{\beta=1}^{K} c(\boldsymbol{n}_\beta)\,\delta(\boldsymbol{n}, \boldsymbol{n}_\beta). \tag{6.3}$$

The weights $c(\boldsymbol{n}_\beta)$ can be interpreted as the volume fraction of fibers oriented in direction $\boldsymbol{n}_\beta$ with respect to the total volume fraction c_F. The relations in Equation (6.2) imply the following properties of $c(\boldsymbol{n}_\beta)$

$$c(\boldsymbol{n}_\beta) \geq 0, \quad \sum_{\beta=1}^{K} c(\boldsymbol{n}_\beta) = 1, \quad c(\boldsymbol{n}_\beta) = c(-\boldsymbol{n}_\beta), \quad \forall \beta = 1 \dots K. \tag{6.4}$$

We further introduce the abbreviation $c_\beta := c\,(\boldsymbol{n}_\beta)$. A uniform empirical fiber orientation distribution can be expressed by uniform weights $c_\beta = 1/K$ and, e.g., a (planar) isotropic distribution of $\boldsymbol{n}_\beta$ on the unit circle or sphere, respectively. The more directions $\boldsymbol{n}_\beta$ being considered, the better a continuous fiber orientation distribution can be approximated by the discrete weights.

6.2.2 Homogenization of linear elastic behavior

The following section describes the estimation of the effective (macroscopic) stiffness tensor and stress localization tensors based on the Mori-Tanaka estimate (Mori and Tanaka, 1973). The relation between the macroscopic stress $\bar{\boldsymbol{\sigma}}$ and strain $\bar{\varepsilon}$ is given by the actual macroscopic stiffness tensor $\bar{\mathbb{C}}$

$$\bar{\boldsymbol{\sigma}} = \bar{\mathbb{C}}\,[\bar{\varepsilon}]\,. \tag{6.5}$$

The fundamental assumption of the Mori-Tanaka homogenization scheme is that the strain localization in the fibers is calculated from the phase-averaged matrix strain ε_{M} instead of the macroscopic strain, by the application of the classical Eshelby (1957) relation. The Mori-Tanaka stiffness thus reads (see, e.g., Brylka (2017); Benveniste (1987))

$$\bar{\mathbb{C}} = \mathbb{C}_{\mathrm{M}} - c_{\mathrm{F}} \left(c_{\mathrm{M}} \left\langle \left(\mathbb{P}_0 + (\mathbb{C}_{\mathrm{F}} - \mathbb{C}_{\mathrm{M}})^{-1} \right)^{-1} \right\rangle_{\mathrm{F}}^{-1} + c_{\mathrm{F}}\, (\mathbb{C}_{\mathrm{F}} - \mathbb{C}_{\mathrm{M}})^{-1} \right)^{-1}. \tag{6.6}$$

An explicit expression of the symmetric polarization tensor $\mathbb{P}_0 = \mathbb{E}_0^{-1}\mathbb{C}_{\mathrm{M}}$ can be found in Ponte Castañeda and Suquet (1998), with $\mathbb{E}_0$ being the Eshelby tensor. An analytical, continuous formulation (following Equation (6.1)) of the orientation average of an arbitrary reference tensor $\mathbb{A}$ over all fibers $\langle\mathbb{A}\rangle_{\mathrm{F}}$ in terms of second and fourth order orientation tensors is given by Advani and Tucker III (1987). A reformulation of the fiber orientation average based on the Rayleigh product and the empirical formulation (see Equation (6.3)) leads to

$$\langle\mathbb{A}\rangle_{\mathrm{F}} = \sum_{\beta=1}^{K} c_\beta \boldsymbol{Q}\left(\boldsymbol{n}_\beta\right) \star \mathbb{A}_0, \quad \boldsymbol{Q}\left(\boldsymbol{n}_\beta\right) \in Orth, \tag{6.7}$$

where $\mathbb{A}_0 = \mathbb{A}\left(\boldsymbol{e}_1\right)$ is an arbitrary tensor in the reference orientation $\boldsymbol{e}_1$ and $\boldsymbol{Q}\left(\boldsymbol{n}_\beta\right)$ is defined as the rotation (i.e. a proper orthogonal tensor) between the reference orientation $\boldsymbol{e}_1$ and $\boldsymbol{n}_\beta$. The phase-averaged matrix and fiber stresses $\boldsymbol{\sigma}_{\mathrm{M}}$ and $\boldsymbol{\sigma}_{\mathrm{F}}$ can be expressed with their corresponding phase-averaged stress localization tensors $\mathbb{B}_{\mathrm{M}}^{\mathrm{MT}}$ and $\mathbb{B}_{\mathrm{F}}^{\mathrm{MT}}$ of the Mori-Tanaka homogenization

$$\boldsymbol{\sigma}_{\mathrm{M}} = \mathbb{B}_{\mathrm{M}}^{\mathrm{MT}}[\overline{\boldsymbol{\sigma}}], \qquad \boldsymbol{\sigma}_{\mathrm{F}} = \mathbb{B}_{\mathrm{F}}^{\mathrm{MT}}[\overline{\boldsymbol{\sigma}}]. \tag{6.8}$$

The localization tensors are determined by

$$\mathbb{B}_{\mathrm{M}}^{\mathrm{MT}} = \left(c_{\mathrm{M}}\mathbb{I}^{\mathrm{s}} + c_{\mathrm{F}}\langle\mathbb{B}_{\mathrm{F0}}^{\mathrm{SIP}}\rangle_{\mathrm{F}}\right)^{-1} \quad \text{and} \quad \mathbb{B}_{\mathrm{F}}^{\mathrm{MT}} = \langle\mathbb{B}_{\mathrm{F0}}^{\mathrm{SIP}}\rangle_{\mathrm{F}}\mathbb{B}_{\mathrm{M}}^{\mathrm{MT}}. \tag{6.9}$$

The fiber stress localization tensor in the single inclusion problem (SIP) in the reference orientation $\mathbb{B}_{\mathrm{F0}}^{\mathrm{SIP}}$ is given by

$$\mathbb{B}_{\mathrm{F0}}^{\mathrm{SIP}} = \left(\mathbb{I}^{\mathrm{s}} + \mathbb{C}_{\mathrm{M}}\left(\mathbb{I}^{\mathrm{s}} - \mathbb{P}_0\mathbb{C}_{\mathrm{M}}\right)\left(\mathbb{C}_{\mathrm{F}}^{-1} - \mathbb{C}_{\mathrm{M}}^{-1}\right)\right)^{-1}. \tag{6.10}$$

Duschlbauer et al. (2003) outlined the calculation of the directional dependent fiber stress $\boldsymbol{\sigma}_{\mathrm{F}}^{\angle}\left(\boldsymbol{n}_\beta\right)$

$$\boldsymbol{\sigma}_{\mathrm{F}}^{\angle}\left(\boldsymbol{n}_\beta\right) = \mathbb{B}_{\mathrm{F}}^{\mathrm{SIP}\angle}\left(\boldsymbol{n}_\beta\right)\mathbb{B}_{\mathrm{M}}^{\mathrm{MT}}[\overline{\boldsymbol{\sigma}}]. \tag{6.11}$$

A rotation of $\mathbb{B}_{\mathrm{F0}}^{\mathrm{SIP}}$ such that the transversely isotropic axis points in direction $\boldsymbol{n}_\beta$ leads to $\mathbb{B}_{\mathrm{F0}}^{\mathrm{SIP}\angle}$

$$\mathbb{B}_{\mathrm{F}}^{\mathrm{SIP}\angle}(\boldsymbol{n}_\beta) = \boldsymbol{Q}(\boldsymbol{n}_\beta) \star \mathbb{B}_{\mathrm{F0}}^{\mathrm{SIP}}, \quad \boldsymbol{Q} \in Orth. \tag{6.12}$$

Figure 6.2 visualizes the directionally-dependent fiber principal stresses for a horizontal, uniaxial tensile load and a planar, isotropic fiber orientation distribution. Arrows indicate the principal stress direction. In all fiber directions, the stress state is almost planar. The fibers in tensile direction experience the highest principal stress. Compressive stresses in fibers perpendicular to the tensile direction arise because the fibers contract less than the matrix in the lateral direction. If the composite strength is loaded on the material combinations considered here, the principal stress in the fiber never reaches the 95% confidence level of the fiber strength. Inter alia, this motivates the assumption of the neglection of fiber breakage.

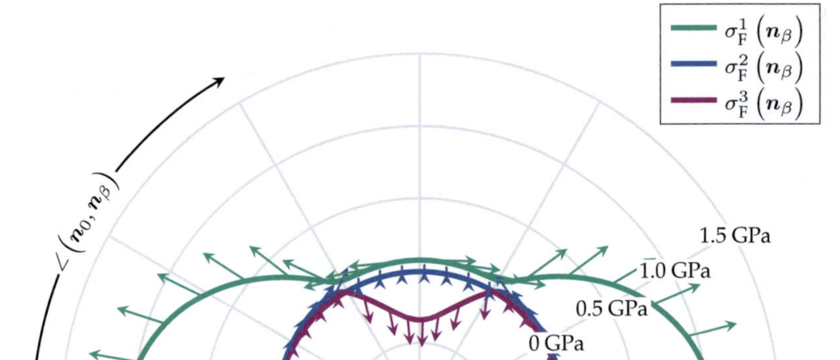

Figure 6.2: Fiber principal stresses (σ_{F}^{1}, σ_{F}^{2}, σ_{F}^{3}) as a function of the fiber orientations $\boldsymbol{n}_\beta$ under macroscopic uniaxial tension in the horizontal direction ($\bar{\boldsymbol{\sigma}} = \bar{\sigma}\boldsymbol{n}_0 \otimes \boldsymbol{n}_0$). Arrows indicate the principal stress direction. The material is UPPH SMC loaded with 130MPa which is 48% of its tensile strength. Damage was not considered in this calculation.

6.2.3 Modeling the matrix damage

The damage behavior of the matrix is modeled by an isotropic degradation of the initial matrix stiffness $\mathbb{C}_{\mathrm{M}}^{0}$ which leads to the following relation for the isotropic matrix stiffness $\mathbb{C}_{\mathrm{M}}$

$$\mathbb{C}_{\mathrm{M}} = (1 - d_{\mathrm{M}})\, \mathbb{C}_{\mathrm{M}}^{0}. \tag{6.13}$$

The damage variable d_{M} is determined as a function of the maximum value of the phase-averaged principal matrix stresses $\sigma_{\mathrm{M}}^{\alpha}$ ($\alpha = 1, 2, 3$) in the prior loading history of the matrix

$$d_{\mathrm{M}} = d_{\mathrm{M}} \left(\max_{\tau \in [0,t]} \left(\max_{\alpha=1,2,3} \sigma_{\mathrm{M}}^{\alpha} \right) \right). \tag{6.14}$$

Since the thermoset matrix is considered a brittle material, we applied a maximum stress criterion asssuming that the material failure is governed by the highest principal stress. The outer max-function ensures that d_{M} is monotonic increasing, i.e., no healing, even for load histories that include unloading.

6.2.4 Modeling the fiber-matrix interface debonding

We consider cylindrical fibers with a large aspect ratio, on which we ignore the debonding of head surfaces (Figure 6.3). By doing so, we firstly assume, that the influence of debonding of the head surfaces on the effective stiffness is small. Secondly, we neglect the coupling of debonding of head surfaces with other damage phenomena, e.g., we ignore crack propagation into the lateral surface or matrix.

We assume that interface damage is governed by the stress on the lateral surface of the interface only. Cauchy's Lemma gives the interface stress vector $\boldsymbol{t}$ as a function of the fiber orientation $\boldsymbol{n}_{\beta}$ and the lateral surface normal $\boldsymbol{s}$ (see Figure 6.3) if the phase-averaged stress tensor $\boldsymbol{\sigma}_{\mathrm{F}}^{\angle}(\boldsymbol{n}_\beta)$ of a fiber with orientation $\boldsymbol{n}_\beta$ is known

$$\boldsymbol{t}(\boldsymbol{n}_\beta, \boldsymbol{s}) = \boldsymbol{\sigma}_{\mathrm{F}}^{\angle}(\boldsymbol{n}_\beta)[\boldsymbol{s}]. \tag{6.15}$$

We decompose the interface stress vector into its normal σ_{I} and shear τ_{I} components thus

$$\boldsymbol{t}(\boldsymbol{n}_\beta, \boldsymbol{s}) = \sigma_{\mathrm{I}}(\boldsymbol{n}_\beta, \boldsymbol{s})\boldsymbol{s} + \tau_{\mathrm{I}}(\boldsymbol{n}_\beta, \boldsymbol{s})\boldsymbol{m}, \ \boldsymbol{s} \perp \boldsymbol{n}_\beta, \ \boldsymbol{m} \perp \boldsymbol{s}, \ \tau_{\mathrm{I}}(\boldsymbol{n}_\beta, \ \boldsymbol{s}) \geq 0, \tag{6.16}$$

where the unit vector $\boldsymbol{m}$ is determined by the tangential part of the interface traction vector $\boldsymbol{t}$ (see Figure 6.3). The component σ_{I} is obtained by a projection of the interface stress vector on $\boldsymbol{s}$, and the component τ_{I}, by the norm of the two tangential projections τ_1 and τ_2

$$\sigma_{\mathrm{I}} = \boldsymbol{t} \cdot \boldsymbol{s}, \tag{6.17}$$

$$\tau_{\mathrm{I}} = \sqrt{\tau_1^2 + \tau_2^2} = \sqrt{(\boldsymbol{t} \cdot \boldsymbol{n}_\beta)^2 + (\boldsymbol{t} \cdot (\boldsymbol{n}_\beta \times \boldsymbol{s}))^2}. \tag{6.18}$$

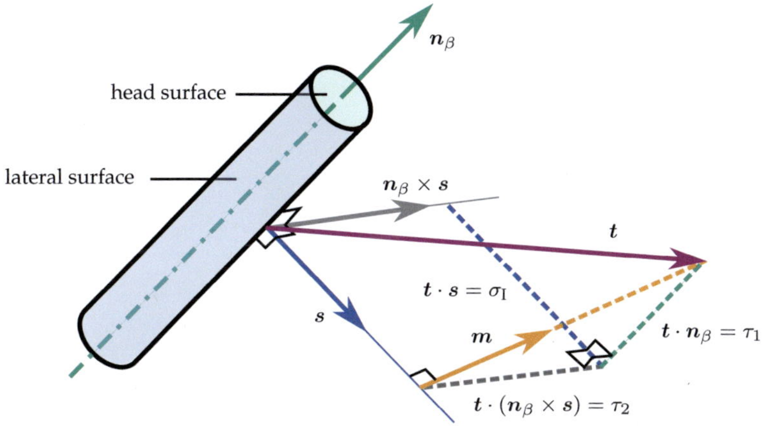

Figure 6.3: Single fiber with orientation $\boldsymbol{n}_\beta$, an outer normal vector $\boldsymbol{s}$ and a tangential vector $\boldsymbol{m}$ on its shell surface.

The maximal interface stress components $\sigma_{\mathrm{I,max}}$ and $\tau_{\mathrm{I,max}}$ are obtained as a function of the fiber direction $\boldsymbol{n}_\beta$ by a maximization over the shell surface

$$\sigma_{\mathrm{I,max}}(\boldsymbol{n}_\beta) = \max_{\boldsymbol{s} \perp \boldsymbol{n}_\beta} \{\sigma_\mathrm{I}(\boldsymbol{n}_\beta, \boldsymbol{s})\}, \tag{6.19}$$

$$\tau_{\mathrm{I,max}}(\boldsymbol{n}_\beta) = \max_{\boldsymbol{s} \perp \boldsymbol{n}_\beta} \tau_\mathrm{I}(\boldsymbol{n}_\beta, \boldsymbol{s}). \tag{6.20}$$

Here, the Macauley bracket $\{x\} = \max(x, 0)$ ensures that only the crack-opening (positive) normal stress is considered.

Figure 6.4 shows the directionally-dependent distribution of the interface stress components on the lateral fiber surface for uniaxial tension in the horizontal direction and a planar isotropic fiber orientation distribution. The maximum shear stress occurs at 45° with respect to the tensile direction; the maximum normal stress, at 90°.

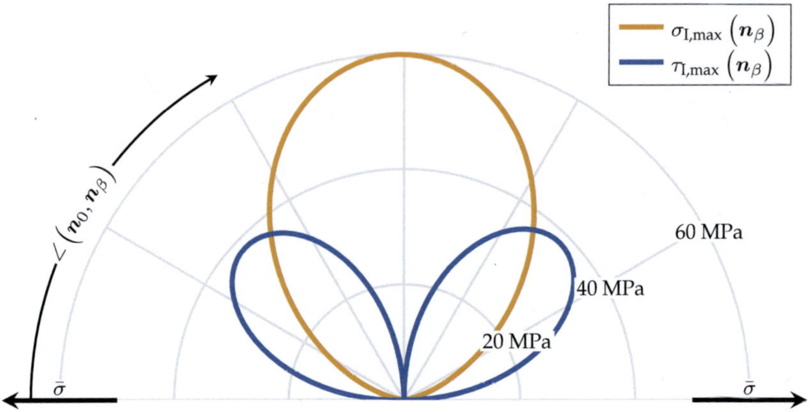

Figure 6.4: Maximum interface shear and normal stress as a function of the fiber orientation relative to a uniaxial tensile load in horizontal direction. The material is UPPH SMC loaded with 130MPa which is 48% of its tensile strength. Damage was not considered in this calculation.

We now introduce an equivalent interface stress $\sigma_{\mathrm{I,eq}}$ that governs interface failure

$$\sigma_{\mathrm{I,eq}}(\boldsymbol{n}_\beta, \boldsymbol{s}) = \hat{\sigma}_{\mathrm{I,eq}} \sqrt[m]{\left(\frac{\tau_{\mathrm{I}}(\boldsymbol{n}_\beta, \boldsymbol{s})}{\tau_{\mathrm{I0}}}\right)^m + \left(\frac{\{\sigma_{\mathrm{I}}(\boldsymbol{n}_\beta, \boldsymbol{s})\}}{\sigma_{\mathrm{I0}}}\right)^m}. \tag{6.21}$$

Three of the four material parameters – $\hat{\sigma}_{\mathrm{I,eq}}$, τ_{I0}, σ_{I0} and m – are independent. Once again, only non-negative normal stress $\{\sigma_{\mathrm{I}}(\boldsymbol{n}_\beta, \boldsymbol{s})\}$ contributes to the equivalent stress.

Under the assumption of a weakest-link failure concept for the interface of a single fiber, a Weibull strength distribution for an inhomogeneous stress field can be adopted (Weibull, 1951). The survival probability $P_{\mathrm{s}}(\boldsymbol{n}_\beta)$ of one fiber in direction $\boldsymbol{n}_\beta$ is given by

$$P_{\mathrm{s}}(\boldsymbol{n}_\beta) = \exp\left(-\frac{1}{A_{\mathrm{I0}}} \int_{A_{\mathrm{I}}} \left\{\frac{\sigma_{\mathrm{I,eq}}(\boldsymbol{n}_\beta, \boldsymbol{s}) - \sigma_{\mathrm{u}}}{\sigma_o}\right\}^k \mathrm{d}A_{\mathrm{I}}\right). \tag{6.22}$$

Here, A_I is the lateral surface area of one fiber. The material parameters A_I0, σ_u, σ_o, and k characterize the strength distribution of the interface. Most models in the literature either consider the anisotropic damage evolution in each interface (Fitoussi et al., 1996a; 1998; Desrumaux et al., 2000) or assume a homogeneous stress distribution on the interface (Ju and Lee, 2000). To the authors' knowledge, no damage model for SMC composites, so far, considers the inhomogeneous stress distribution on the interface in a weakest-link approach on each interface yet. This leads to a simplified formulation of the survival probability $P_\mathrm{s,max}$

$$P_\mathrm{s,max}\left(\boldsymbol{n}_\beta\right) = \exp\left(-\frac{A_\mathrm{I}}{A_\mathrm{I0}}\left\{\frac{\sigma_\mathrm{I,eq}\left(\boldsymbol{n}_\beta, \boldsymbol{s}_\mathrm{max}\right) - \sigma_\mathrm{u}}{\sigma_o}\right\}^k\right). \tag{6.23}$$

Figure 6.5 shows a typical survivial probability under the assumption of a homogeneous stress distribution on the interface.

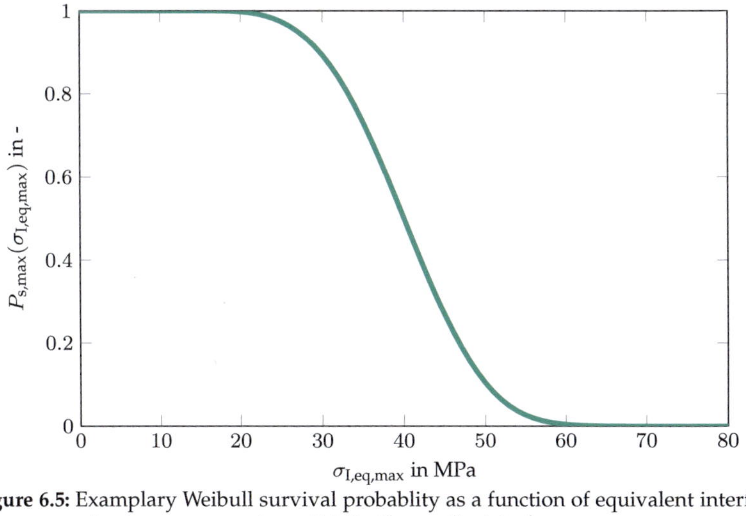

Figure 6.5: Examplary Weibull survival probablity as a function of equivalent interface stress, assuming homogenous stress distribution on the interface.

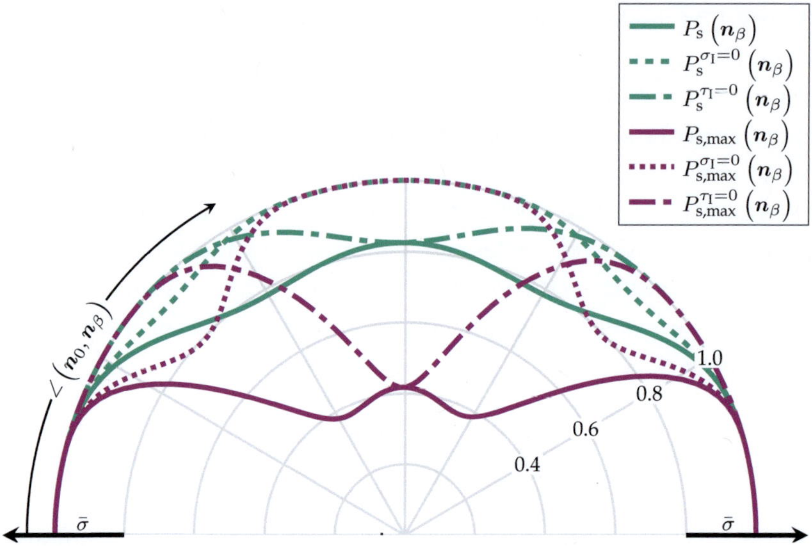

Figure 6.6: Comparison of survival probability approximations and different special cases, assuming an inhomogeneous interface stress distribution (P_s), and a homogeneous distribution of the maximal interface stress ($P_{s,max}$). The material is UPPH SMC loaded with 120MPa which is 44% of its tensile strength. Damage was not considered in this calculation.

Figure 6.6 compares the two failure criteria P_s and $P_{s,max}$. It also shows their behavior when excluding shear stress $P_s^{\tau_I=0}$, $P_{s,max}^{\tau_I=0}$ and when excluding normal stress $P_s^{\sigma_I=0}$, $P_{s,max}^{\sigma_I=0}$. Both criteria have equal regions of no chance of failure ($P_s = 1$, $P_{s,max} = 1$). However, the failure probability under the assumption of a homogeneous stress distribution on the interface is significantly higher. Hereafter, P_s is applied for the estimation of the survival probability.

We now assume that interfaces can be damaged and thus lead to a loss of load-carrying capability. For each fiber orientation $\boldsymbol{n}_\beta$, the initial distribution of load-carrying fibers c_β^0 equals the fiber orientation distribution c_β. The fiber fraction c_β can be decomposed into the fraction of intact load-carrying fibers $\bar{c}_\beta$ and damaged fibers $\breve{c}_\beta$

$$c_\beta = \bar{c}_\beta + \breve{c}_\beta. \tag{6.24}$$

For a damaged material, the variable $\bar{c}_\beta$ does not represent a normed fiber orientation distribution ($\sum_{\beta=1}^{K} \bar{c}_\beta < 1$).
The fraction of intact fibers in direction $\boldsymbol{n}_\beta$ with respect to the initial fraction of fibers in that direction is given by the probability $P_\mathrm{i}(\boldsymbol{n}_\beta)$ of finding a load-carrying fiber in the corresponding direction

$$P_\mathrm{i}(\boldsymbol{n}_\beta) = \begin{cases} 1, & c_\beta = 0, \\ \dfrac{\bar{c}_\beta}{c_\beta}, & c_\beta > 0. \end{cases} \tag{6.25}$$

We state that the probability $P_\mathrm{i}(\boldsymbol{n}_\beta)$ of finding intact fibers in direction $\boldsymbol{n}_\beta$ can be no higher than the probability $P_\mathrm{s}(\boldsymbol{n}_\beta)$ of fibers in that direction surviving a given external load, represented by $\sigma_{\mathrm{I,eq}}(\boldsymbol{n}_\beta, \boldsymbol{s})$. This key assumption leads to the following direction-dependent damage function and consistency condition

$$\phi(\boldsymbol{n}_\beta) = P_\mathrm{i}(\boldsymbol{n}_\beta) - P_\mathrm{s}(\boldsymbol{n}_\beta) \overset{!}{\leq} 0. \tag{6.26}$$

Equation (6.26) leads to a natural evolution for the orientation distribution of load-carrying fibers. For simplicity, we assume that fibers with damaged interfaces behave like matrix material. This behavior can be modeled by an artificial increase in the matrix volume fraction ($\dot{c}_\mathrm{M} \geq 0$), and a decrease in the fiber volume fraction ($\dot{c}_\mathrm{F} \leq 0$). The latter one is given by

$$c_\mathrm{F} = c_\mathrm{F}^0 \sum_{\beta=1}^{K} \bar{c}_\beta. \tag{6.27}$$

Remark:

In case of the simplified maximum interface stress formulation of the survival probability $P_{\text{s,max}}\left(\sigma_{\text{I,eq}}\left(\boldsymbol{n}_\beta, \boldsymbol{s}_{\text{max}}\right)\right)$ we can derive an evolution equation for $\dot{\bar{c}}_\beta$ in analogy to classic plasticity models. For simplicity, we introduce the maximum stress on the interface in direction $\boldsymbol{n}_\beta$ as $\sigma_{\text{I,eq}}^{\text{max}}\left(\boldsymbol{n}_\beta\right) := \sigma_{\text{I,eq}}\left(\boldsymbol{n}_\beta, \boldsymbol{s}_{\text{max}}\right)$. Starting with Equation (6.26), we introduce the Lagrange parameters $\lambda_\beta\ (\beta = 1 \ldots K)$ and obtain an evolution equation of the intact fiber fraction $\bar{c}_\beta$ in rate form

$$\dot{\bar{c}}_\beta = -\lambda_\beta \frac{\partial \phi\left(\boldsymbol{n}_\beta, \sigma_{\text{I,eq}}^{\text{max}}\left(\boldsymbol{n}_\beta\right)\right)}{\partial \sigma_{\text{I,eq}}^{\text{max}}\left(\boldsymbol{n}_\beta\right)} \quad \forall \beta = 1 \ldots K. \tag{6.28}$$

Here, two possibilities exist: i) The deformation yields no interface damage in direction $\boldsymbol{n}_\beta$, $\phi\left(\boldsymbol{n}_\beta, \sigma_{\text{I,eq}}^{\text{max}}\left(\boldsymbol{n}_\beta\right)\right) \leq 0$ holds and $\dot{\bar{c}}_\beta = 0$, therefore $\lambda_\beta = 0$ since $\partial \phi\left(\boldsymbol{n}_\beta, \sigma_{\text{I,eq}}^{\text{max}}\left(\boldsymbol{n}_\beta\right)\right) / \partial \sigma_{\text{I,eq}}^{\text{max}}\left(\boldsymbol{n}_\beta\right) > 0$. ii) The deformation yields interface damage in direction $\boldsymbol{n}_\beta$, $\phi\left(\boldsymbol{n}_\beta, \sigma_{\text{I,eq}}^{\text{max}}\left(\boldsymbol{n}_\beta\right)\right) = 0$ holds and $\dot{\bar{c}}_\beta < 0$, therefore $\lambda_\beta > 0$. These constraints can also be formulated equivalently as Karush-Kuhn-Tucker conditions (Karush, 1939; Kuhn and Tucker, 1951)

$$\phi\left(\boldsymbol{n}_\beta, \sigma_{\text{I,eq}}^{\text{max}}\left(\boldsymbol{n}_\beta\right)\right) \leq 0, \quad \lambda_\beta \geq 0, \quad \forall \beta = 1 \ldots K, \tag{6.29}$$

$$\lambda_\beta \phi\left(\boldsymbol{n}_\beta, \sigma_{\text{I,eq}}^{\text{max}}\left(\boldsymbol{n}_\beta\right)\right) = 0, \quad \forall \beta = 1 \ldots K. \tag{6.30}$$

In the discrete formulations (Section 6.3), both, Equation (6.26) and Equation (6.29) lead to an algebraic system of equations. Hereby, the calculations of the increments $\Delta\dot{\bar{c}}_\beta$ and λ_β only differ by a factor of $-\partial \phi\left(\boldsymbol{n}_\beta, \sigma_{\text{I,eq}}^{\text{max}}\left(\boldsymbol{n}_\beta\right)\right) / \partial \sigma_{\text{I,eq}}^{\text{max}}\left(\boldsymbol{n}_\beta\right)$, which needs to be calculated. The introduction of a potential relation according to Equation (6.29) in, e.g., a von-Mises plasticity model reduces the system of equations which needs to be solved to a single scalar equation. In constrast to that, here the system of equations remains unchanged by the introduction of Equation (6.29) (K unknowns, K directions). In the following, the focus, therefore, lies on the computation of the current fraction of intact fibers directly based on Equation (6.26).

6.3 Numerical implementation

6.3.1 Incremental formulation

Implicit time integration

We implemented the model as a UMAT (User Material) in the commercial implicit finite element-software ABAQUS/Standard (Simulia Dassault Systèmes, 2014). Thereby, the algorithmic tangent was calculated numerically by a perturbation of all stress components. Our goal was a computationally efficient and robust implementation that allows application to larger structures. Hereafter, the vector $\underline{\bar{c}}$ lists all fractions of intact fibers $\bar{c}_\beta$

$$\underline{\bar{c}} = \begin{pmatrix} \bar{c}_1 \\ \vdots \\ \bar{c}_m \end{pmatrix}. \tag{6.31}$$

The implicit Euler time integration scheme is applied. Here, a tuple of unknowns $\underline{y}^{n+1}$ at time t^{n+1} is calculated using information from this new time step

$$\underline{y}^{n+1} = \underline{y}^{n} + \triangle t \underline{\dot{y}}^{n+1} = \underline{y}^{n} + \triangle t \underline{f}\left(\underline{y}^{n+1}, t^{n+1}\right). \tag{6.32}$$

Because the desired solution $\underline{y}^{n+1}$ appears on both sides of the equation, an implicit algebraic equation must be solved. This is quite compatible with the algebraic formulation resulting from Equation (6.26), as shown in the following.

Application to the model

The discretization and implementation of the matrix damage (Equation (6.14)) is straight forward. Application of the implicit Euler time integration yields

$$d_{\mathrm{M}}^{n+1} = d_{\mathrm{M}}^{n+1}\left(\max_{\tau\in[0,t]}\left(\max_{\alpha=1,2,3}\sigma_{\mathrm{M}}^{\alpha,n+1}\right)\right). \tag{6.33}$$

The maximum, absolute principal matrix stress $\sigma_{\mathrm{M}}^{\alpha,n+1}$ at time t^{n+1} is calulated from the matrix stress $\boldsymbol{\sigma}_{\mathrm{M}}^{n+1}$ following Equation (6.8). It implicitly depends on all current fractions of intact fibers $\underline{\bar{c}}^{n+1}$.
The intact fiber fractions in the current time step t^{n+1} are

$$\underline{\bar{c}}^{n+1} = \underline{\bar{c}}^{n} + \triangle t\dot{\underline{\bar{c}}}^{n+1}. \tag{6.34}$$

A corresponding evolution equation for $\dot{\underline{\bar{c}}}^{n+1}$ will be presented in Section 6.3.2.

Resulting system of equations

The resulting system of equations to be solved comprises all consistency conditions (Equation (6.26)) with active damage directions $\boldsymbol{d}_\alpha$ ($\alpha = \{1 \dots L\}$, where L is the number of active directions) and the matrix damage equation (Equation (6.14)). Substituting relations (6.33) and (6.34) yields the nonlinear system of equations

$$\begin{aligned} \phi\left(\boldsymbol{d}_1, \underline{\bar{c}}^{n+1}, d_{\mathrm{M}}^{n+1}\right) &\stackrel{!}{=} 0 \\ &\vdots \\ \phi\left(\boldsymbol{d}_L, \underline{\bar{c}}^{n+1}, d_{\mathrm{M}}^{n+1}\right) &\stackrel{!}{=} 0 \\ d_{\mathrm{M}}^{n+1} = d_{\mathrm{M}}^{n+1}&\left(\max_{\tau\in[0,t]}\left(\max_{\alpha=1,2,3}\sigma_{\mathrm{M}}^{\alpha,n+1}\right)\right) \end{aligned} \tag{6.35}$$

which must be solved for the unknown entities $\underline{\bar{c}}^{n+1}$ and d_{M}^{n+1}. Due to the equivalent interface stress (Equation (6.21)) which is based on the Mori-Tanaka localization (Equation (6.8)), the resulting equation system (Equation (6.35)) is nonlinear in the intact fiber fractions $\underline{\bar{c}}$ and the matrix damage d_{M}. Identification of active damage directions $\boldsymbol{d}_\alpha$ will be discussed in Section 6.3.2.

6.3.2 Solving the system of equations

The consistency condition (Equation (6.26)) yields an evolution equation for the fraction of intact fibers $\bar{c}_\beta$ in direction $\boldsymbol{n}_\beta$. Consider $q \leq K$ directions in which the consistency condition (Equation (6.26)) is not fulfilled. An active damage direction $\boldsymbol{d}_\alpha$ is estimated by the direction with the highest violation of the consistency condition (Equation (6.26))

$$\boldsymbol{d}_\alpha = \underset{\boldsymbol{n}_\beta \forall \beta \in [1,q]}{\operatorname{argmax}} \left(\phi\left(\boldsymbol{n}_\beta, \underline{\bar{c}}, d_{\mathrm{M}}\right)\right). \tag{6.36}$$

The intact fiber fractions in the current time step t^{n+1} (Equation (6.34)) are modeled as

$$\underline{\bar{c}}^{n+1} = \underline{\bar{c}}^{n} + \sum_{\alpha=1}^{L} \gamma_\alpha \underline{\underline{R}}_\alpha \underline{\bar{c}}^{0}. \tag{6.37}$$

Here, L is the number of active damage directions, γ_α is a Lagrange coefficient representing the share of newly damaged fibers in the corresponding direction, $\underline{\underline{R}}_\alpha$ is a regularization matrix, and $\underline{\bar{c}}^0$ are the initial intact fiber fractions. The regularization matrix smears the initially Dirac-like damage in a single direction to neighboring fiber directions, according to the definition

$$\underline{\underline{R}}_\alpha = \operatorname{diag}\left(\cos^p\left(\angle\left(\boldsymbol{d}_\alpha, \boldsymbol{n}_1\right)\right), \cos^p\left(\angle\left(\boldsymbol{d}_\alpha, \boldsymbol{n}_2\right)\right), \ldots, \cos^p\left(\angle\left(\boldsymbol{d}_\alpha, \boldsymbol{n}_K\right)\right)\right). \tag{6.38}$$

The parameter $p \in [1, \infty)$ defines the sharpness of the regularization. The introduction of this kind of regularization is based on the redistribution of load to neighboring directions as soon as interfaces in a considered direction are damaged (see Figure 6.9a). To circumvent expensive calculations, the suggested regularization can be applied in order to distribute beforehand the damage in a certain range around a considered direction. Figure 6.7 shows regularizations of a single Dirac-like horizontal damage direction $\boldsymbol{d}_1$ for different parameters p. Hereby, a regularization parameter of $p \to \infty$ means that regularization is not considered. Small values of the regularization parameter ($p \to 1$) correspond to a high regularization. According to Figure 6.7 and Equation (6.38), respectively, the share of directions which are significantly affected by a single active damage direction through the regularization, increases with a decrease of p.

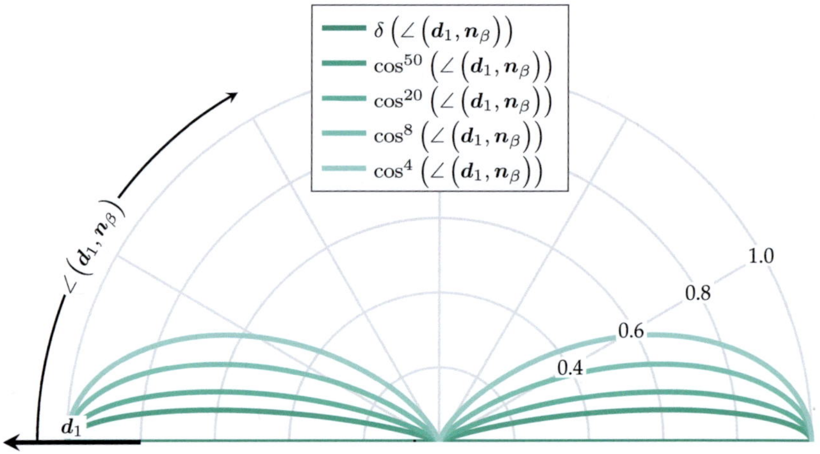

Figure 6.7: Regularization of the horizontal damage direction $\boldsymbol{d}_1$ for different regularization parameters p.

The size of the system of equations (Equation (6.35)) that needs to be solved, i.e., the number of active damage directions L, can therefore be reduced significantly by a suitable regularization. In doing so, numerical stability and efficiency are both increased.

According to Equation (6.37), the unknowns to be solved for are γ_α and d_{M}^{n+1}. The resulting solution $(\underline{\bar{c}}^{n+1}, d_{\mathrm{M}}^{n+1})$ needs to fulfill all consistency conditions. If no convergence is achieved yet, the solution is considered incorrect, and an additional active damage direction $\boldsymbol{d}_\alpha$ according to Equation (6.36) needs to be taken into account. This additional active damage direction for the next iteration is chosen as that direction from the current iteration with the highest violation of the consistency condition. The size of the equation system (6.35) is thus increased by one. This straight forward "active set search" of a single newly-considered active damage direction $\boldsymbol{d}_\alpha$ in each iteration leads to a comparatively small equation system (see, e.g., Miehe (1996)). This system then needs to be solved to gain a solution $(\underline{\bar{c}}^{n+1}, d_{\mathrm{M}}^{n+1})$, such that all consistency conditions are fulfilled. Detailed information on the numerical solution of constrained nonlinear algebraic equations and active sets can be found in Shacham (1986) and Nocedal and Wright (2006). We applied a Banach fixed-point iteration and a Quasi-Newton scheme.

The essential aspects of the implemented SMC composite damage model are shown in the flow chart in Figure 6.8.

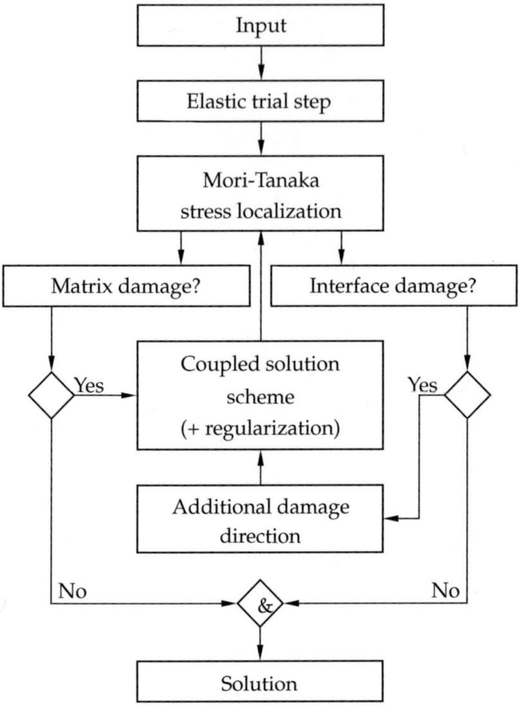

Figure 6.8: Flow chart of the implemented SMC composite damage model.

Figure 6.9 shows the solutions for different regularization parameters p in a uniaxial stress state. The left column shows γ_α as a function of the macroscopic strain. The planar isotropic fiber orientation distribution is discretized in $K = 179$ directions on the half circle. Therefore, a maximal number of $\alpha \in (L \leq K = 179)$ damage directions $\boldsymbol{d}_\alpha$ are possible. Each point indicates an active damage direction $\boldsymbol{d}_\alpha$ at a specific strain level, that is, the number of points on a vertical line corresponds to the number of active damage directions. The point color illustrates the value of γ_α. The more reddish a point, the higher the share of fibers with newly debonded interfaces in the corresponding time step. The tensile direction in the left column is the $\boldsymbol{n}_1$ direction ($\alpha = 1$) which is the lowest abscissa axis. The right column shows a polar plot of an approximation of the load-carrying directional fiber content $\bar{c}_\beta$, evaluated at strain levels corresponding to the left column. The tensile direction in the right column lies on the horizontal axis. The first row (Figure 6.9a and Figure 6.9b) was obtained without regularization ($p \to \infty$) and, therefore, exhibits a sharp damage initiation in each direction. In each strain step, the full equation system ($L = K = 179$) is solved. After a certain applied strain level (respectively load level), most of the interfaces in a direction are debonded and, therefore, the value of γ_α in this direction decreases. Simultaneously, the load is spread over neighboring directions and the value of γ_α in these directions increases. The second row (Figure 6.9c and Figure 6.9d) is the result for a regularization of $p = 20$. The deviation of $\bar{c}_\beta$ and the macroscopic behavior between the first and second row is low. The regularization in the third row (Figure 6.9e and Figure 6.9f) is relatively high ($p = 8$). The evolution of $\bar{c}_\beta$ shows significant deviations relative to the non-regularized solution. Only three ($L = 3$) or fewer damage directions are active in each time step.

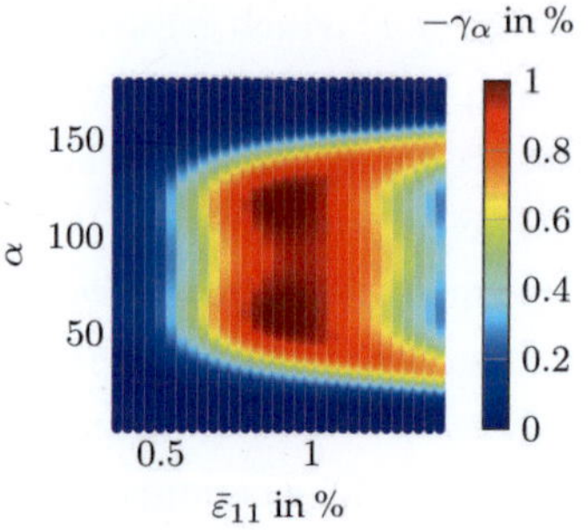

(a) Evolution of γ_α for $p \to \infty$

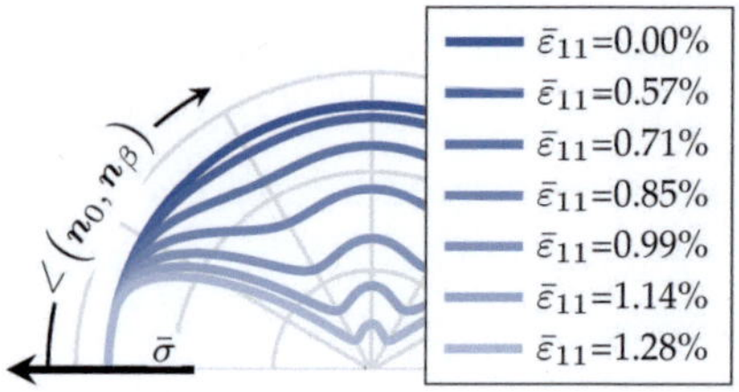

(b) Evolution of $\bar{c}_\beta$ for $p \to \infty$

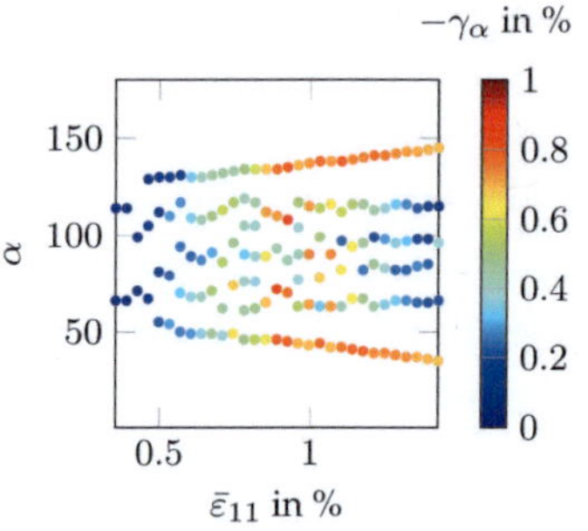

(c) Evolution of γ_α for $p = 20$

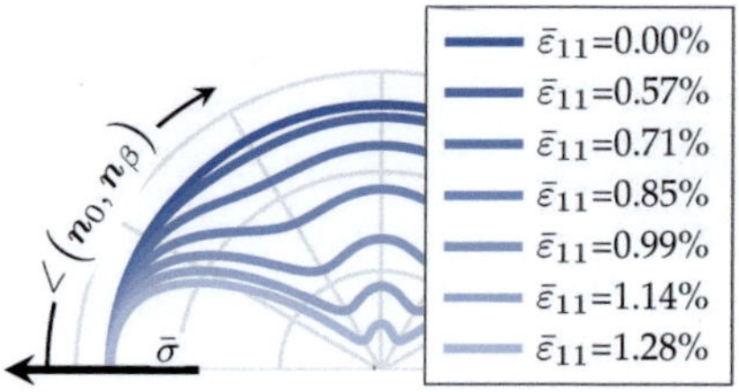

(d) Evolution of $\bar{c}_\beta$ for $p = 20$

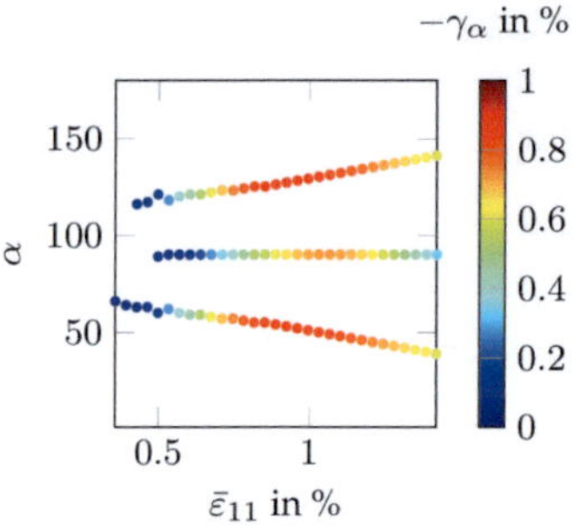

(e) Evolution of γ_α for $p = 8$

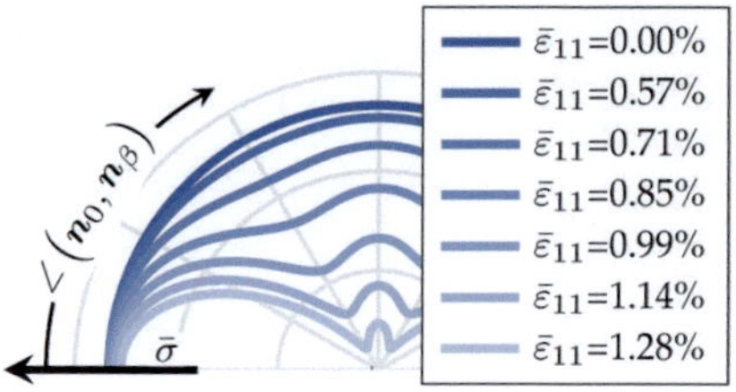

(f) Evolution of $\bar{c}_\beta$ for $p = 8$

Figure 6.9: Evolution of γ_α and $\bar{c}_\beta$ for different regularizations $\underline{\underline{R}}_\alpha$ under uniaxial tension in the $\boldsymbol{n}_1$ direction. The tensile direction corresponds to the horizontal direction in the right polar plots. The initially planar isotropic fiber orientation distribution is discretized in $K = 179$ directions on the half circle.

6.4 Parameter identification

6.4.1 Matrix damage

In this chapter, two matrix materials are considered: epoxy resin (reinforced with 43vol.% and 50vol.% e-glass fibers) and UPPH resin (reinforced with 23vol.% e-glass fibers). Monotonic tensile tests of neat resin bone specimens were performed to characterize the matrix behavior. The epoxy neat resin samples were casted pressureless in a net shape mold. The UPPH neat resin samples were manufactured by a project partner in the International Research Training Group GRK 2078 (Trauth, 2017). Figure 6.10 shows the stress-strain behavior and the evolution of d_M under the assumption that the nonlinear behavior arises solely from damage (i.e., nonlinear viscoelasticity and plasticity are neglected). For the computations, $d_M(\varepsilon_M)$ was approximated by an ansatz such that d_M is zero until a damage initiation strain threshold followed by a, sufficiently precise fit with a ninth-order polynomial for d_M above the damage initiation threshold.

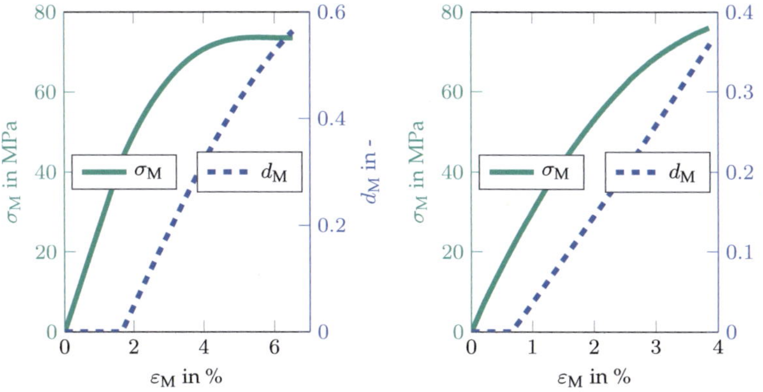

(a) Tensile behavior of epoxy

(b) Tensile behavior of UPPH

Figure 6.10: Tensile behavior of the matrix systems.

For instance, Kinloch and Young (1983) and Morgan and O'Neal (1977) reported microcracks in neat epoxy under tensile load which mainly result in stiffness degradation. The estimation of d_{M} from solely the secant modulus is, therefore, considered to be sufficient. The determination of irreversible stiffness degradation is, in general, challenging when not only considering damage but a general nonlinear elasto-viscoplastic damage material behavior (Brylka et al., 2018). The strain rate for the matrix samples was set to the estimated average matrix strain rate experienced in the composite tensile tests. Thereby, linear viscous effects can be excluded.

6.4.2 Fiber-matrix interface strength distribution

The fiber and matrix material properties, the fiber surface properties, the roving composition, the fiber sizing, and the process-dependent fiber impregnation characteristics all influence the interface strength distribution. The interface failure behavior is not directly characterized in this work. Instead, we obtain the interface properties from the literature and inverse parameter identification with tensile tests performed on the SMC composite.

The literature commonly applies a linear criterion ($m = 1$, Equation (6.21)) (Fitoussi et al., 1996a;b; 1998; Derrien et al., 2000; Desrumaux et al., 2000; 2001; Meraghni et al., 2002), a quadratic criterion ($m = 2$) (Ben Cheikh Larbi et al., 2006; Tandon et al., 2002; Koyanagi et al., 2012; Fitoussi et al., 1996b; Jendli et al., 2009; Ogihara and Koyanagi, 2010; Swentek, 2014), or a parabolic criterion (cubic contribution of shear stress and quadratic contribution of normal stress) (Koyanagi et al., 2012; Ogihara and Koyanagi, 2010). Typical experimental results (Koyanagi et al., 2012; Ogihara and Koyanagi, 2010) fit similarly to the quadratic and parabolic criterion. Here, the more commonly considered quadratic criterion (Tandon et al., 2002; Koyanagi et al., 2012; Ogihara and Koyanagi, 2010; Swentek, 2014) is applied. The contributions of the normal and shear stress on the failure behavior can be quantified by $\sigma_{\mathrm{I0}}/\tau_{\mathrm{I0}}$. Experimentally, this relation was obtained in a lap shear test by Swentek (2014) ($\sigma_{\mathrm{I0}}/\tau_{\mathrm{I0}} \approx 1.6$). Ogihara and Koyanagi (2010) ($\sigma_{\mathrm{I0}}/\tau_{\mathrm{I0}} = 1.3 - 1.8$) and Tandon et al. (2002) ($\sigma_{\mathrm{I0}}/\tau_{\mathrm{I0}} \approx 1.22$) measured the interface strength of epoxy and glass fibers using a fragmentation test on cruciform specimens with varying angles between the specimens' arms. Koyanagi et al. (2012) ($\sigma_{\mathrm{I0}}/\tau_{\mathrm{I0}} = 1.3$) compared the method to single-fiber pull-out tests under a combined stress state. In the following, we assume $\sigma_{\mathrm{I0}}/\tau_{\mathrm{I0}} = 1.5$.

The parameters $\hat{\sigma}_{\mathrm{I,eq}}, \tau_{\mathrm{I0}}, \sigma_{\mathrm{I0}}, A_{\mathrm{I0}}, \sigma_{\mathrm{u}}, \sigma_{\mathrm{o}}$ and k describe the interface strength. If $\sigma_{\mathrm{I0}}/\tau_{\mathrm{I0}}$ is given, three of the five remaining parameters are independent. Following Equation (6.21) and Equation (6.22), these independent parameters ξ_1, ξ_2 and ξ_3 are

$$\begin{aligned} \xi_1 &:= \frac{1}{(A_{\mathrm{I0}})^{1/k}} \frac{1}{\sigma_{\mathrm{o}}} \hat{\sigma}_{\mathrm{I,eq}} \frac{\tau_{\mathrm{I0}}}{\sigma_{\mathrm{I0}}}, \\ \xi_2 &:= \frac{1}{(A_{\mathrm{I0}})^{1/k}} \frac{\sigma_{\mathrm{u}}}{\sigma_{\mathrm{o}}}, \\ \xi_3 &:= k. \end{aligned}$$

We estimate these three parameters by fitting the model to uniaxial tensile tests on bone specimens. We assumed an initially planar isotropic microstructure, since the flow lengths in the compression molding were short. Figure 6.11 shows the fit and experimental results for the epoxy SMC and UPPH SMC with the low fiber volume fractions.The resulting interface strength distributions are plotted in Figure 6.12 under the assumption of a homogeneous interface stress distribution. Interpreting the interface strength distribution requires caution, since it does not capture the real distribution, but instead, also serves as an implicit model corrector. For example, the assumption of homogeneous matrix damage underestimates the composite stiffness degradation. Thus, the fitting procedure results in a lower interface strength distribution than the real distribution.

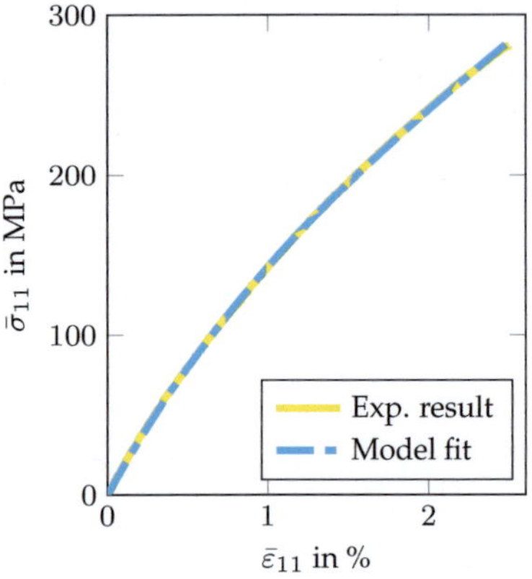

(a) epoxy matrix, $c_{\mathrm{F}}^{0} = 43\%$

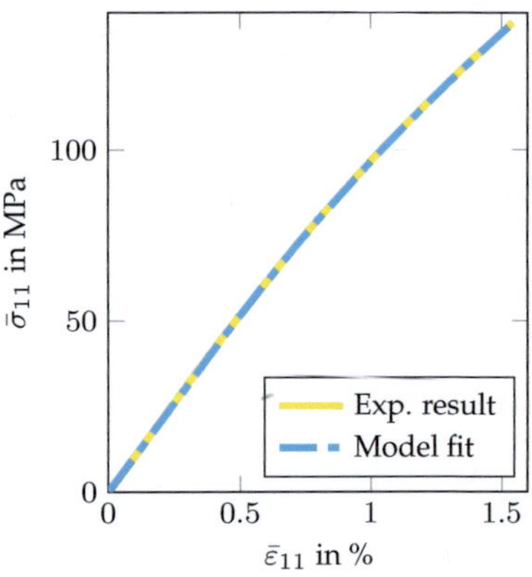

(b) UPPH matrix, $c_{\mathrm{F}}^{0} = 23\%$

Figure 6.11: Experimental tensile test results (dotted lines) with the corresponding model fits (solid lines).

The fitted interface strengths (Figure 6.12) show a significantly wider range than the interface strength typically measured in the literature (e.g., Broutman (1969)), the shape of which is similar to the strength distribution in Figure 6.5. Such a more narrow strength distribution results in a stress-strain behavior of approximately bilinear nature.

6.5 Application

6.5.1 Variation of fiber content

The epoxy matrix SMC was available with two fiber contents. After fitting the model to the lower fiber content ($c_{\mathrm{F}}^{0} = 43\%$), we applied the model to the higher fiber content ($c_{\mathrm{F}}^{0} = 50\%$) under the assumption that the interface strength distribution is not affected by the fiber content and, thus, remains constant. The simulations were performed in one macroscopic homogeneous material point.

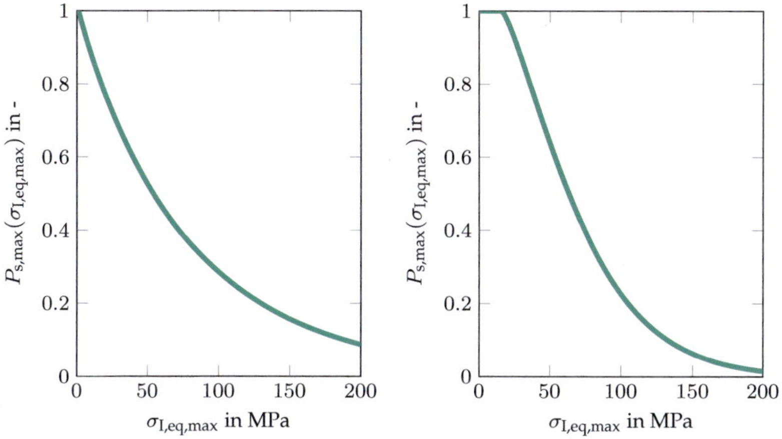

(a) epoxy matrix, $c_F^0 = 43\%$ **(b)** UPPH matrix, $c_F^0 = 23\%$

Figure 6.12: Interface survival probabilities (assumption: homogenous stress distribution on the interface), resulting from fits in Figure 6.11.

Figure 6.13a shows the simulated and measured stress-strain behavior. The model slightly over-estimates the stiffness reduction for an increasing fiber volume fraction. Figure 6.13b depicts the estimated evolution of the total load-carrying fiber volume fraction c_F and the relative matrix stiffness reduction d_M. The matrix degradation is underestimated by the phase-averaged isotropic matrix damage model. As the fiber volume fraction increases, the matrix volume fraction decreases, respectively, and with it, the influence of the underestimated matrix damage decreases, as well. The higher influence of the interface damage model, therefore, results in an overestimation of the overall stiffness reduction.

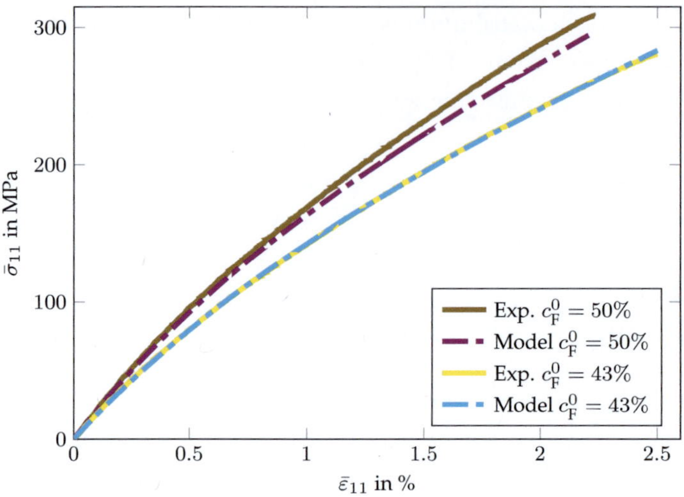

(a) Stress-strain behavior

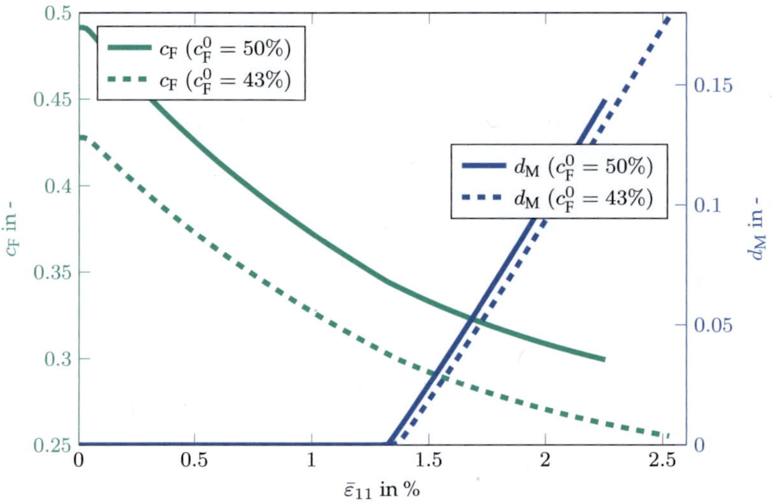

(b) Evolution of c_F and d_M

Figure 6.13: Validation of epoxy SMC under uniaxial tension for two different volume fraction (interface strength distribution fitted to $c_F^0 = 43\%$).

Another source of the deviation between the experiment and the model prediction might be a fiber-content-dependent interface strength distribution. One possible explanation for an increase in the interface strength with increasing fiber content might be the improved fiber filament impregnation caused by roving dispersion during fiber-fiber interaction in the manufacturing process.

6.5.2 Biaxial tensile loading

Cruciform specimens that allow for comparably high biaxial and homogeneous stress made it possible to validate the damage model in a biaxial stress state. The specimens arms were reinforced with continuous tapes manufactured in a co-molding process. A detailed discussion of the cruciform design and experimental procedure will be covered in an upcoming publication. These samples were only available for UPPH SMC which have a B-stage (partially cured resin) that allows for co-molding with UD-reinforcements.

Figure 6.14a compares the model prediction and the experimental results. Figure 6.14b displays the estimated evolution of total load-carrying fiber volume fraction c_{F} and matrix damage d_{M}. The evolution of interface and matrix damage is earlier and faster under biaxial tension, compared to uniaxial tension. The failure strain is significantly lower under biaxial tension (0.95% vs. 1.54%). The tensile strength, likewise, is lower under biaxial than under uniaxial tension ($111\mathrm{MPa}$ vs. $136\mathrm{MPa}$).

The model predicts the stress under biaxial tension with a maximal relative error of about 4%. This result is satisfactory, considering the typically high fluctuations in the mechanical properties of SMC.

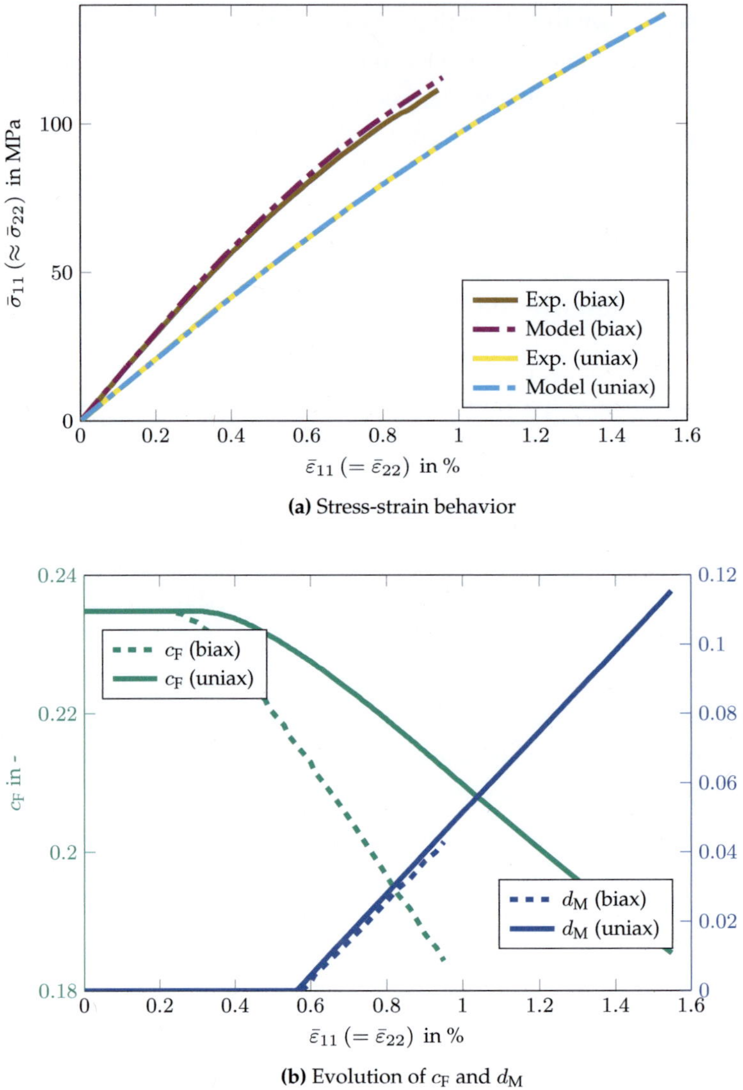

(a) Stress-strain behavior

(b) Evolution of c_F and d_M

Figure 6.14: Validation of UPPH SMC under a uniaxial and biaxial stress state (interface strength distribution fitted to uniaxial stress state).

6.5.3 Application to different load paths

Here, we present detailed simulation results of additional load cases of the UPPH SMC with a fiber volume fraction of $c_{\mathrm{F}}^{0} = 23\%$. We did not validate these load cases experimentally. Rather, the numerical results serve to illustrate the performance of the two-scale damage model for non-proportional loading scenarios. The thereby considered ultimate loadings somewhat exceed the strength of the material. All load cases consist of two subsequent loading steps, applied at a constant strain rate in a plane stress state. The directionally Young's modulus was calculated following the procedure of Böhlke and Brüggemann (2001).

The first load case (in detail documented in Figure 6.15) is given by uniaxial compression in the $\boldsymbol{e}_1$-direction followed by uniaxial tension in the $\boldsymbol{e}_1$-direction, respectively. In the first loading step, the interface shear damage, dominantly $45°$ to the tensile direction, governs the damage behavior. The interface damage due to normal stress (which is highest perpendicular to the tensile direction) evolves only in the second loading step. As we modeled matrix damage in dependence of the highest principal matrix stress, matrix damage also mainly evolves in the second loading step.

The second load case (in detail documented in Figure 6.16) is tension in $\boldsymbol{e}_1$-direction and subsequent tension in $\boldsymbol{e}_2$-direction. In the first loading step, the stiffness reduction is perpendicular to the tensile direction, that is the $\boldsymbol{e}_2$-direction, which mainly derives from interface damage. Accordingly, the initial stiffness in the second loading step is significantly lower. The interface damage due to shear stress and matrix damage do not evolve in the second loading step, as their behavior is invariant to this load rotation. The final stiffness is approximately planar isotropic in the shown $\boldsymbol{e}_1$-$\boldsymbol{e}_2$-plane.

The third load case (in detail documented in Figure 6.17) prescribes subsequent planar shear in opposing direction ($\pm\boldsymbol{e}_1$-$\boldsymbol{e}_2$-directions). In the first loading step, the main interface damage appears in the e_1- and e_2-directions. The main damage already evolves during the first loading step. The anisotropy of interface damage with respect to the vertical axis is again due to the fact that we only consider positive normal interface stresses. In the second loading step, the damage evolution is significantly lower. The tensile load leads to a symmetrization of the FODF with respect to the vertical axis visualized by the weights $\bar{c}_\beta$. The small amount of matrix damage at the end of the second loading step is due to the evolution of load partitioning implied by the micromechanical damage model.

The fourth load case (in detail documented in Figure 6.18) is planar shear (e_1-e_2-plane) followed by uniaxial tension in the e_1-direction. The first loading step is, therefore, equal to the first step in the previous load case (Figure 6.17). In the horizontal tension, mainly the fibers in the upper right direction experience interface damage, as their strength is significantly lower than the others (because less interfaces are damaged).

Obviously, the model is able to handle various load cases and loading sequences and yields reasonable results. Experimental validation, however, remains a challenging task.

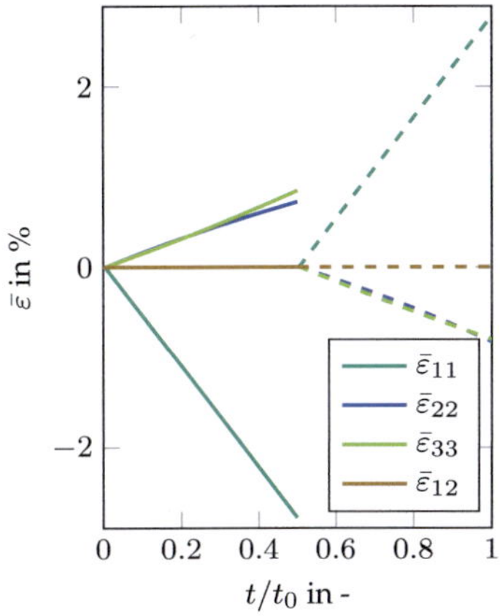

(a) Strain components

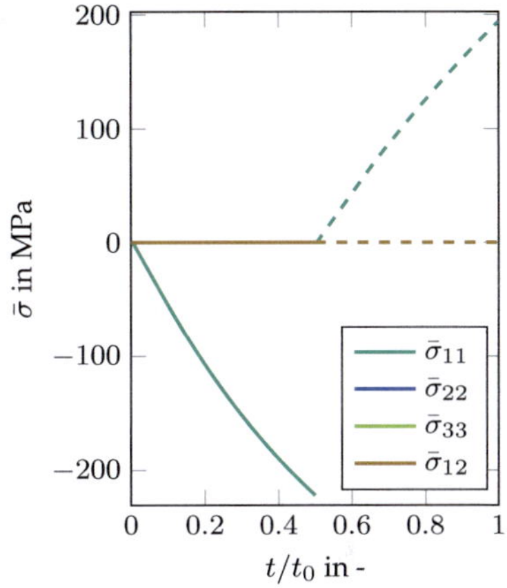

(b) Stress components

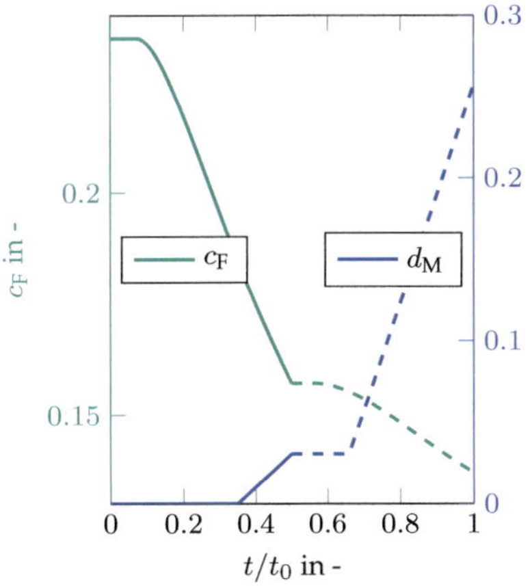

(c) Evolution of load-carrying fiber volume fraction c_F and matrix damage variable d_M

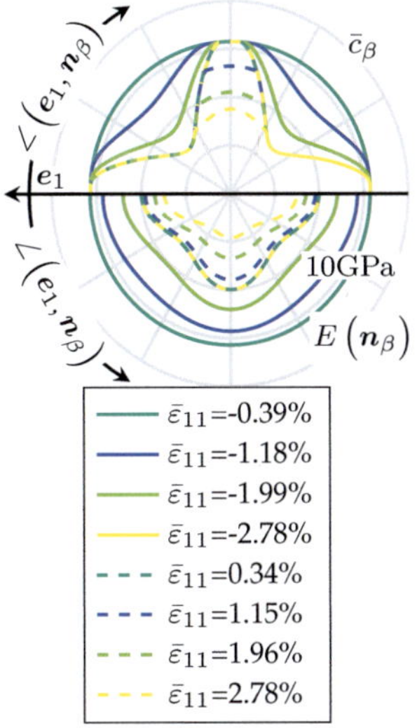

(d) Weights of fiber orientation distribution $\bar{c}_\beta$ (upper half circle) and directional-dependent Young's modulus $E(\boldsymbol{n}_\beta)$ (lower half circle)

Figure 6.15: Model predictions for a subsequent compression (in $\boldsymbol{e}_1$-direction) and tensile (also in $\boldsymbol{e}_1$-direction) load.

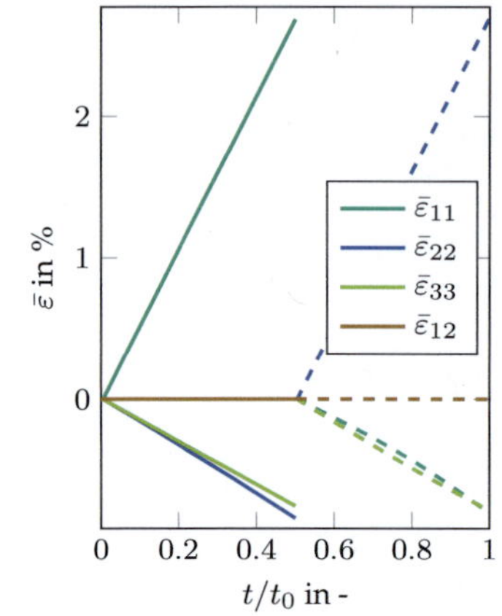

(a) Strain components

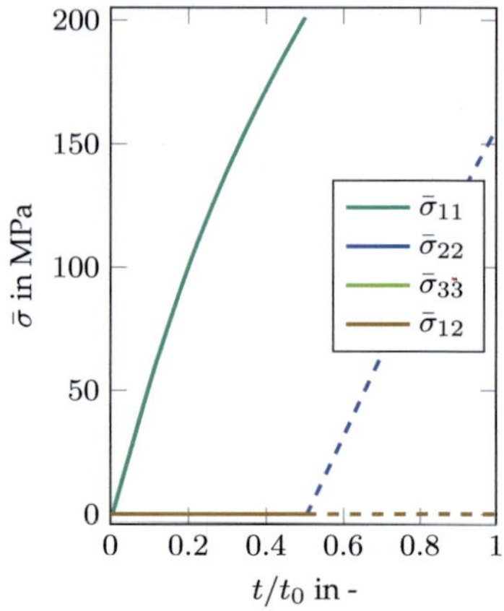

(b) Stress components

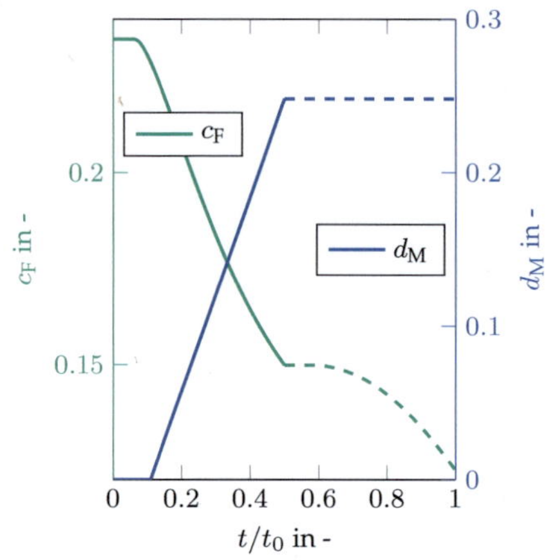

(c) Evolution of load-carrying fiber volume fraction c_F and matrix damage variable d_M

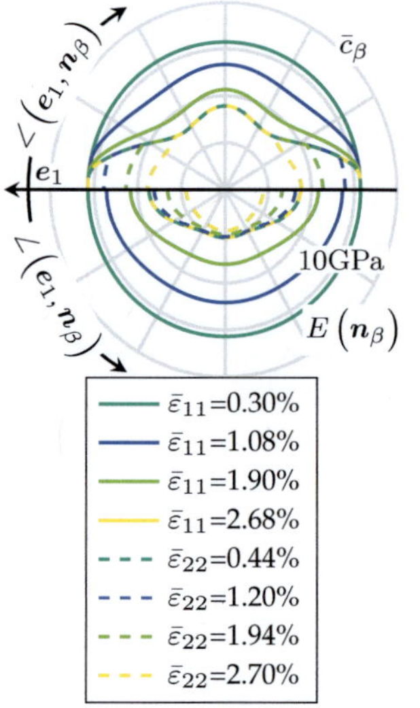

(d) Weights of fiber orientation distribution $\bar{c}_\beta$ (upper half circle) and directional-dependent Young's modulus $E(\boldsymbol{n}_\beta)$ (lower half circle)

Figure 6.16: Model predictions for a subsequent tension in $\boldsymbol{e}_1$-direction and tension in $\boldsymbol{e}_2$-direction.

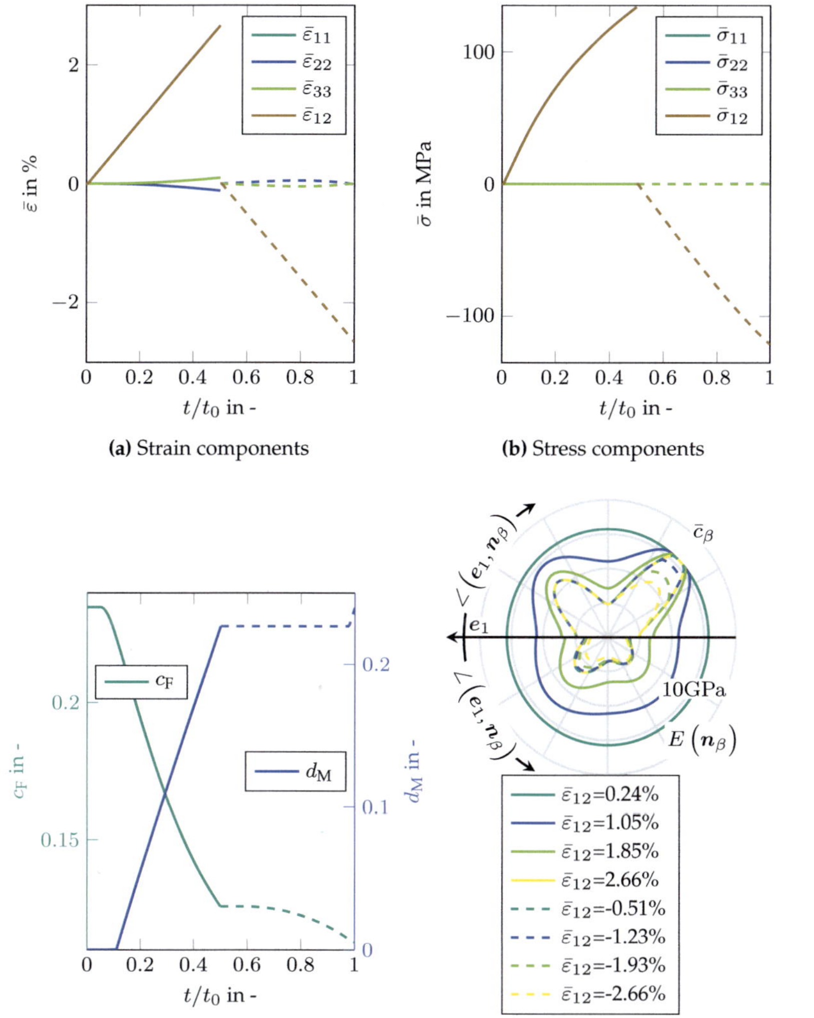

(a) Strain components

(b) Stress components

(c) Evolution of load-carrying fiber volume fraction c_F and matrix damage variable d_M

(d) Weights of fiber orientation distribution $\bar{c}_\beta$ (upper half circle) and directional-dependent Young's modulus $E(\boldsymbol{n}_\beta)$ (lower half circle)

Figure 6.17: Model predictions for a subsequent shear load $\boldsymbol{e}_1$-$\boldsymbol{e}_2$-direction and shear load in negative $\boldsymbol{e}_1$-$\boldsymbol{e}_2$-direction.

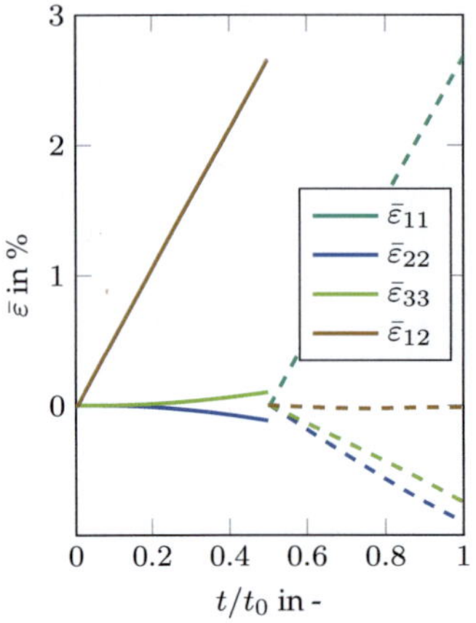

(a) Strain components

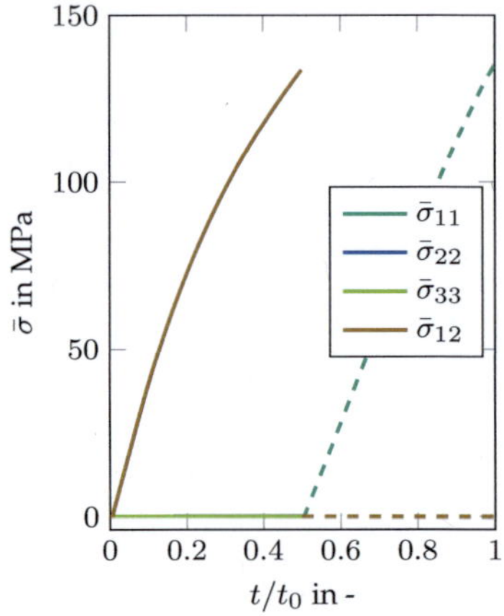

(b) Stress components

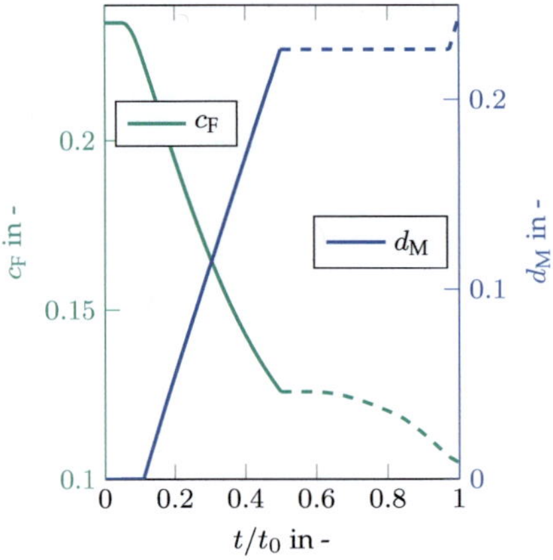

(c) Evolution of load-carrying fiber volume fraction c_F and matrix damage variable d_M

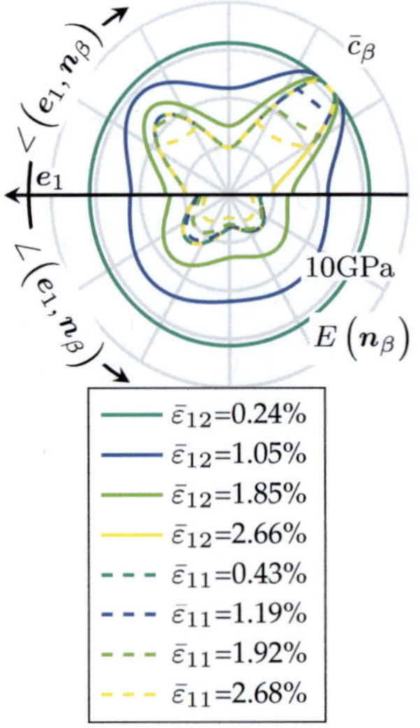

(d) Weights of fiber orientation distribution $\bar{c}_\beta$ (upper half circle) and directional-dependent Young's modulus $E(\boldsymbol{n}_\beta)$ (lower half circle)

Figure 6.18: Model predictions for a subsequent shear load $\boldsymbol{e}_1$-$\boldsymbol{e}_2$-direction and tensile load in $\boldsymbol{e}_1$-direction.

6.6 Conclusions

Our elasto-damage model for SMC captures matrix damage and interface debonding on the microscale. A Mori-Tanaka homogenization scheme is applied to calculate the corresponding macroscopic behavior. The model accounts for an arbitrary, mold-flow-induced, inhomogeneous fiber orientation distribution of straight fibers. The complete model can predict SMC damage behavior for different matrix systems, fiber contents, and stress states. However, the applicability is so far only demonstrated for a small number of material combinations. Further validation is required to evaluate the model capabilities and limitations more precisely.

Matrix damage is modeled as an isotropic degradation of the matrix stiffness based on the maximum principal matrix stress. This approach underestimates the stiffness degradation and does not adequately capture anisotropy due to microcracks. Interface debonding is modeled as a reduction of load-carrying fiber fraction in the directions exposed to sufficiently large equivalent interface stresses. Here, a Weibull interface strength distribution is assumed. We show that an approach that only takes account of the maximum equivalent stress occurring on the interface, notably underestimates the interface survival probability. We, therefore, consider the inhomogeneous stress distribution around the transverse fiber axis. The interface damage model presented here, leads to anisotropic stiffness degradation.

Matrix and interface damage are coupled by the localization relation which does not capture such phenomena as crack propagation from the matrix into the interface or vice versa. Artificial regularization of the considered damage directions and a computationally efficient model implementation permit application to larger components. Characterizing the process-dependent interface strength distribution remains a challenge – particularly for fiber filaments constituted in rovings.

Manufacturing-induced eigenstresses such as those resulting from chemical and thermal shrinkage, influence matrix and interface behavior.

Chapter 7

Summary, conclusions, outlook

In this chapter, the thesis is summarized, chapter by chapter. Then the individual chapters are linked and their correlation is discussed before ending with a general outlook.

Chapter 4 introduces a novel scheme to characterize the anisotropic elastic and macroscopically inhomogeneous SMC in biaxial tensile tests. μCT scans suggest that macroscopic inhomogeneities result mainly from an inhomogeneous fiber orientation distribution. The identification of a generally inhomogeneous stiffness leads, however, to an ill-posed problem allowing for no unique solution. A linear correlation between the stiffness tensor and the fiber orientation distribution is, therefore, assumed. This simplification reduces the problem size to five degrees of freedom per specimen, which do not depend on the fiber orientation distribution. Four of these parameters are identifiable and determined in a Gauss-Newton type optimization procedure.

Chapter 5 presents an investigation of different cruciform specimen designs for damage characterization in SMC. Here, various (material-specific) requirements for optimal specimen design are defined. One key challenge was achieving a high strain level in the center region of the cruciform specimen, while preventing premature failure in the clamped specimen arms. Starting from the ISO norm for sheet metals, design variations are introduced, including two concepts to reinforce the specimens' arms. An experimental evaluation included two different loading scenarios: uniaxial tension and equi-biaxial tension. The best fit, in terms of the defined optimality criteria, was a specimen manufactured in a layup with unidirectional reinforcing outer layers. A gentle milling process exposed the pure SMC in the center region of the specimen. This cruciform specimen performed excellently for all loading conditions considered. For example, in the uniaxial loading scenario, the average strain in the center region reached 87% of the failure strain in a uniaxial tensile bone specimen.

Chapter 6 presents an anisotropic, micromechanical damage model for SMC. The model captures the dominant damage mechanisms – matrix damage and fiber-matrix interface debonding – in a Mori-Tanaka homogenization framework. The matrix damage was modeled as a phase-averaged isotropic stiffness degradation. The interface damage is governed by an equivalent interface stress on the lateral fiber surface. The inhomogeneous stress distribution in the fiber-matrix interface is taken into account in the definition of the equivalent stress governing the fiber-matrix interface damage. A Weibull distribution for the interface strength is assumed. The model can account for anisotropic distribution and evolution of load-carrying fibers with intact interfaces. The model was validated with tensile tests on unsaturated polyester polyurethane hybrid and epoxy resin systems having different glass fiber contents (23-50vol.%). The model yielded satisfyingly accurate predictions under uniaxial and biaxial stress states. However, the applicability is so far only demonstrated for a small number of material combinations. Further validation is required to evaluate the model capabilities and limitations more precisely.

For both the characterization and modeling of SMC, it is important to take the microstructure into account. In Chapter 4, with prior knowledge of the fiber orientation tensor and a simple constitutive assumption, the degrees of freedom could be reduced from infinity to five. These five parameters define the anisotropic elasticity for any fiber orientation distribution. For the biaxial characterization of the damage behavior

(Chapter 5), the inhomogeneity and anisotropy of SMC by testing specimens with various microstructures and orientations needs to be investigated. In the presented damage model (Chapter 6) the SMC microstructure is a key input for estimating the macroscopic stiffness and stress localization in the matrix and on the interface. The model can predict the nonlinear behavior of SMC for any given fiber volume fraction and fiber orientation distribution.

The validation of the damage model with a specimen developed in Chapter 5 under biaxial tension highlights the synergies of this research project. To better understand the anisotropic damage behavior of SMC, further experimental results would be useful, for instance, adding more loading ratios and observing stiffness degradation outside the main loading direction. A microstructure-based parameter identification similar to the one presented in Chapter 4 could be applied to validate or calibrate the model. Additionally, biaxial tensile testing could be used to characterize the interface or matrix behavior.

A visionary application of this or an improved version of the mean-field damage model would consist of a complete, virtual process chain. Mold-filling analysis could provide the fiber orientation distribution, and this result could be validated by μCT analysis. Chemo-thermo-mechanical full-field simulations could give insight into eigenstresses on the microscale and macroscale. A two-scale structural simulation of an SMC component with a mean-field damage model could predict warpage, component stiffness, and strength. Loading experiments on the component level could serve to validate the damage model and the entire virtual process chain.

Appendix A

Specific numerical fiber orientation tensors

This Appendix lists a selection of fiber orientation tensors of the microstructure shown in Figure 4.1. Pinter (2017) captured the image and performed the fiber orientation analysis (Pinter et al., 2018). We list the fiber orientation tensors of the first kind and fourth order $\mathbb{N}$ in the tupel $\underline{\underline{N}}$ according to the defined standard in GRK 2078 (Priesnitz, 2016)

$$\begin{aligned}\underline{\underline{N}} = (&N_{3333}, N_{3332}, N_{3322}, N_{3222}, N_{2222},\\ &N_{3331}, N_{3321}, N_{3221}, N_{2221}, N_{3311},\\ &N_{3211}, N_{2211}, N_{3111}, N_{2111}, N_{1111})^{\mathsf{T}}.\end{aligned}$$

The fiber orientation tensors at points A, B, C, and D in Figure 4.1 are listed in the following

$$\begin{aligned}\underline{\underline{N}}^{\mathrm{A}} \approx (&0.0009, 0.0003, 0.0038, 0.0030, 0.1803,\\ &0.0034, 0.0013, 0.0158, 0.0126, 0.0150,\\ &0.0060, 0.1216, 0.0759, 0.0351, 0.5380)^{\mathsf{T}},\end{aligned}$$

$$\underline{\underline{N}}^{\mathrm{B}} \approx (0.0011, 0.0000, 0.0075, 0.0008, 0.4775,\\ -0.0030, -0.0003, -0.0191, -0.0330, 0.0102,\\ 0.0020, 0.1331, -0.0383, -0.0324, 0.2199)^{\mathsf{T}},$$

$$\underline{\underline{N}}^{\mathrm{C}} \approx (0.0011, -0.0005, 0.0028, -0.0050, 0.0958,\\ 0.0042, -0.0020, 0.0138, -0.0303, 0.0182,\\ -0.0099, 0.1094, 0.0916, -0.0361, 0.6424)^{\mathsf{T}},$$

$$\underline{\underline{N}}^{\mathrm{D}} \approx (0.0021, 0.0007, 0.0097, 0.0053, 0.4480,\\ -0.0049, -0.0016, -0.0234, 0.0069, 0.0142,\\ 0.0032, 0.1301, -0.0483, -0.0040, 0.2420)^{\mathsf{T}}.$$

Appendix B

Strain fields shortly before failure and images of failed cruciform specimens

This appendix shows the strain fields in the area of interest in terms of the last captured image before specimen failure. For better comparison, all contour plots refer to the legend in Figure B.1. Additionally, we present images of the failed specimens. The fractured areas are marked in green.

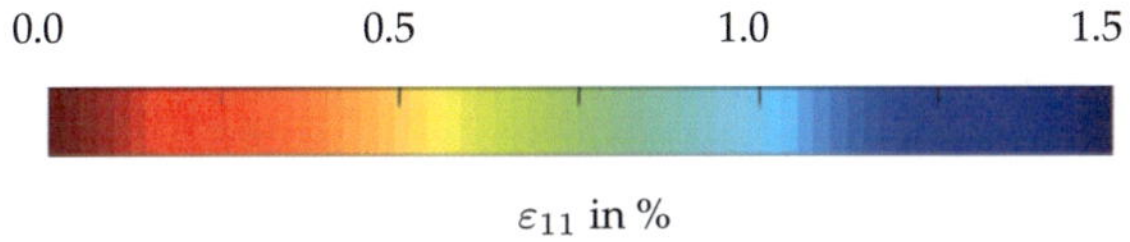

Figure B.1: Legend for the following contour plots

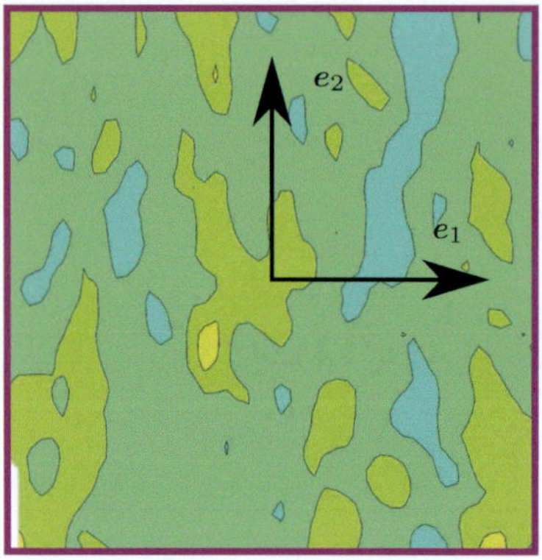

(a) Strain field ε_{11} in A^{I} for $P_{\Gamma=0}$

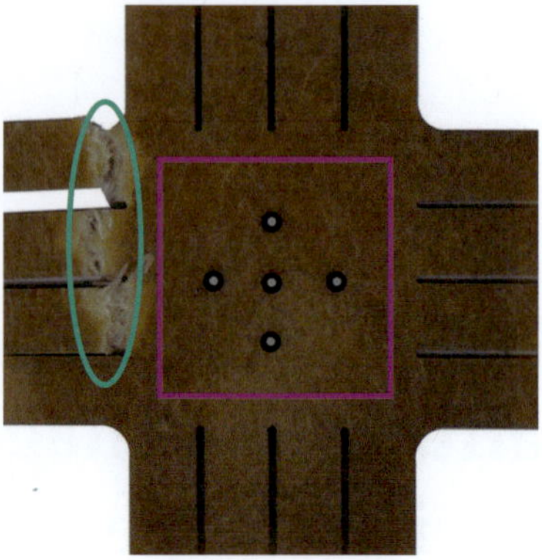

(b) Failed specimen for load case $P_{\Gamma=0}$

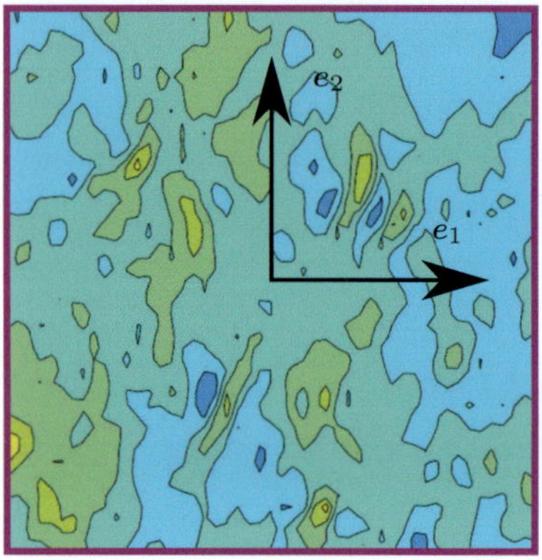

(c) Strain field ε_{11} in A^{I} for $P_{\Gamma=1}$

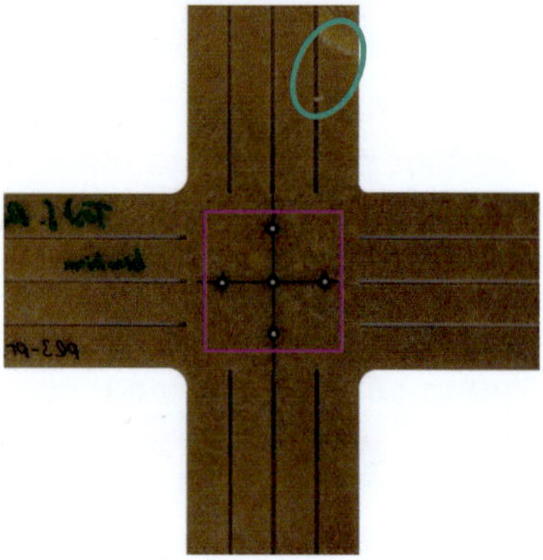

(d) Failed specimen for load case $P_{\Gamma=1}$

Figure B.2: Unreinforced specimen arms. Last image of strain field before failure and failed specimen.

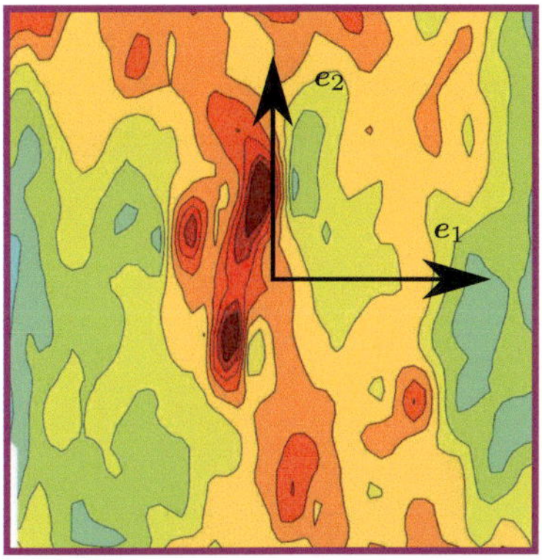

(a) Strain field ε_{11} in A^{I} for $P_{\Gamma=0}$

(b) Failed specimen for load case $P_{\Gamma=0}$

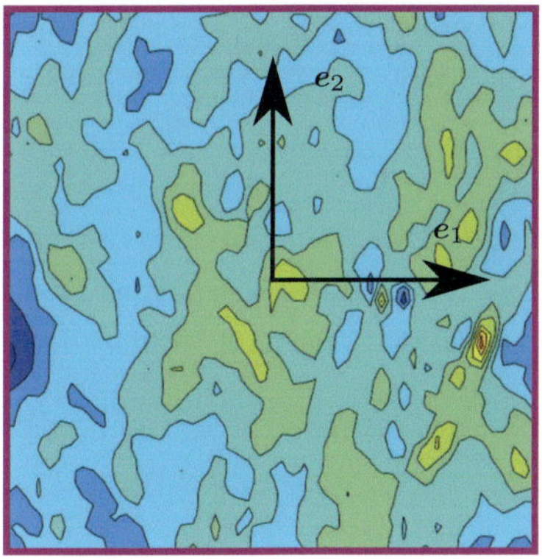

(c) Strain field ε_{11} in A^{I} for $P_{\Gamma=1}$

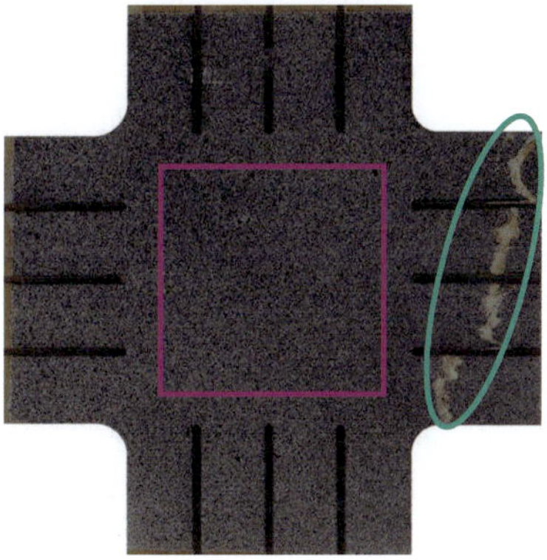

(d) Failed specimen for load case $P_{\Gamma=1}$

Figure B.3: Bonded specimen arms. Last image of strain field before failure and failed specimen.

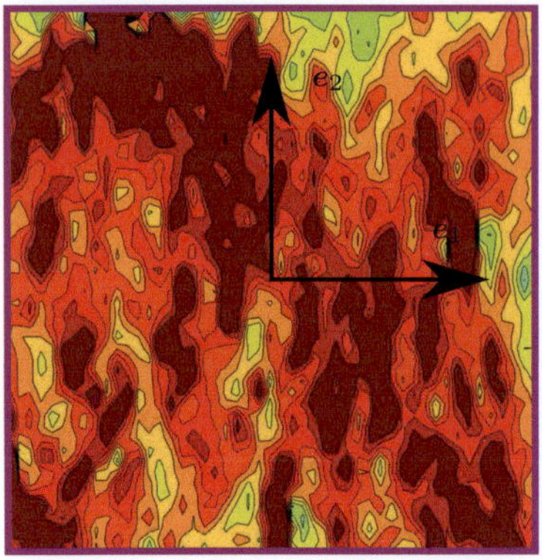

(a) Strain field ε_{11} in A^{I} for $P_{\Gamma=0}$

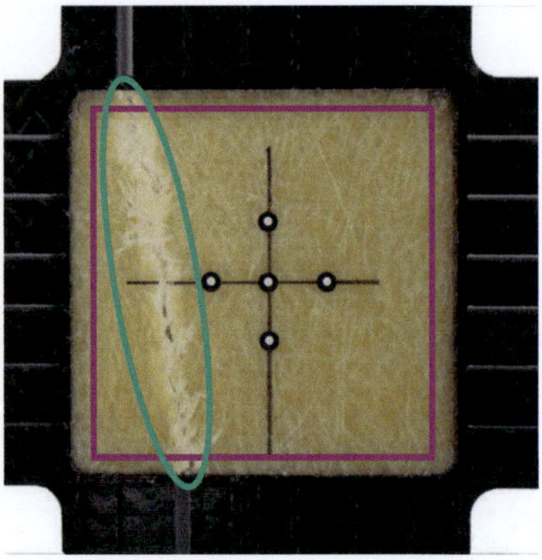

(b) Failed specimen for load case $P_{\Gamma=0}$

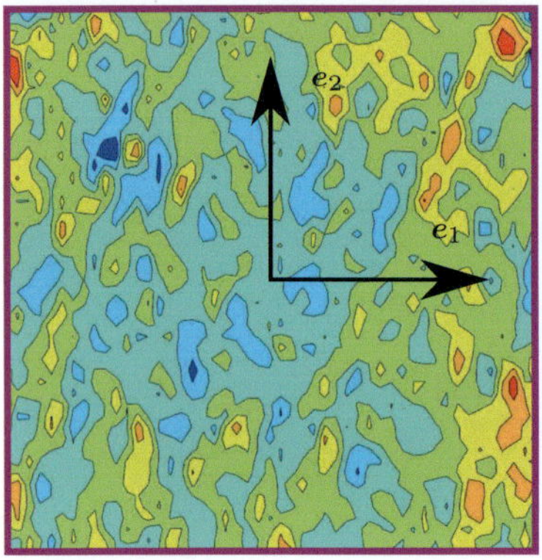

(c) Strain field ε_{11} in A^{I} for $P_{\Gamma=1}$

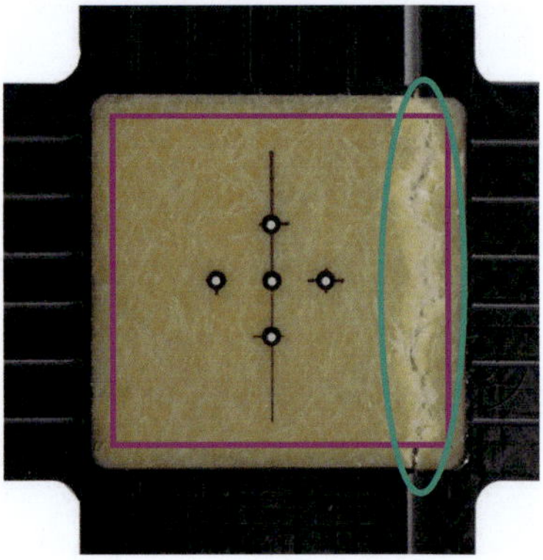

(d) Failed specimen for load case $P_{\Gamma=1}$

Figure B.4: Continuous fiber specimen arms, geometry 1. Last image of strain field before failure and failed specimen.

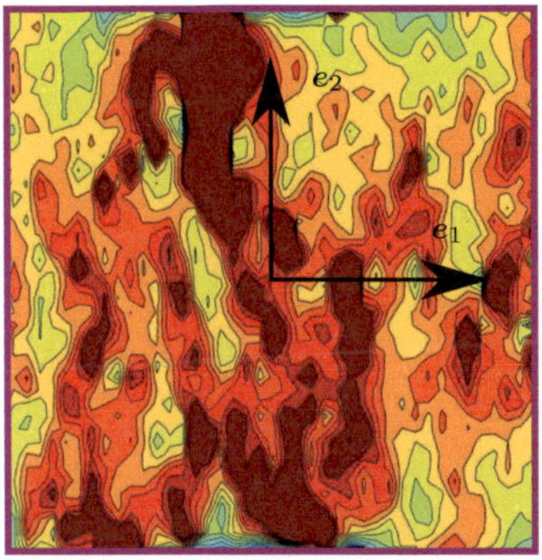

(a) Strain field ε_{11} in A^{I} for $P_{\Gamma=0}$

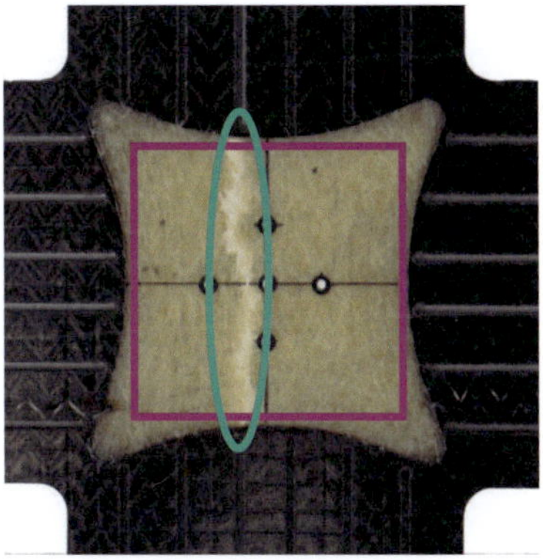

(b) Failed specimen for load case $P_{\Gamma=0}$

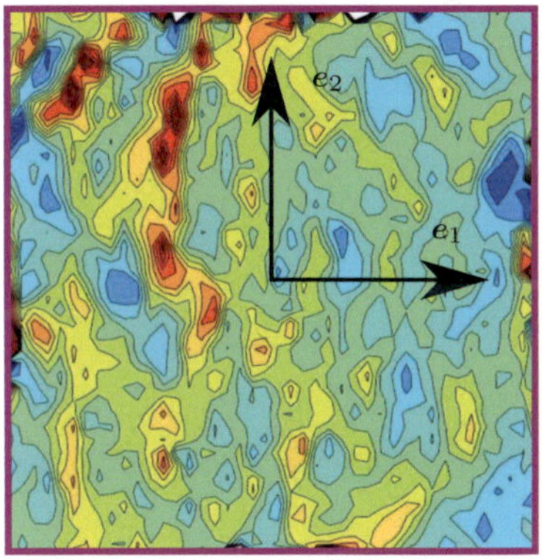

(c) Strain field ε_{11} in A^{I} for $P_{\Gamma=1}$

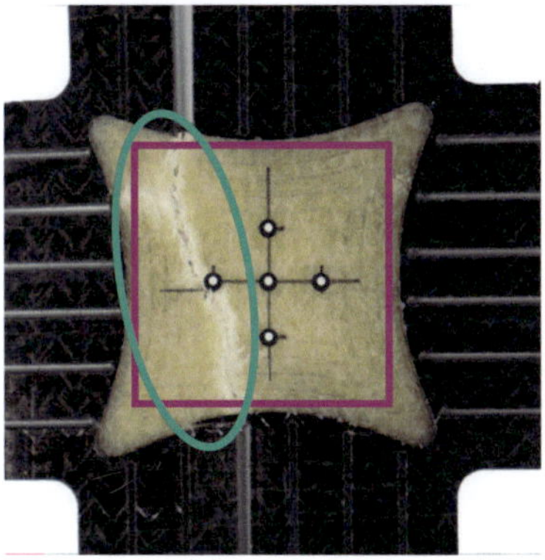

(d) Failed specimen for load case $P_{\Gamma=1}$

Figure B.5: Continuous fiber specimen arms, geometry 2. Last image of strain field before failure and failed specimen.

Frequently used acronyms, symbols, and operators

Acronyms

μCT	Micro-computed tomography
CoDiCoFRP	Discontinuous fiber-reinforced polymer with continuous fiber reinforcements
CoDiCoFRTP	Discontinuous fiber-reinforced thermoplast with continuous fiber reinforcements
CoDiCoFRTS	Discontinuous fiber-reinforced thermoset with continuous fiber reinforcements
CoFRP	Continuous fiber-reinforced polymer
CoFRTP	Continuous fiber-reinforced thermoplast
CoFRTS	Continuous fiber-reinforced thermoset
DIC	Digital image correlation
DiCoFRP	Discontinuous fiber-reinforced polymer
DiCoFRTP	Discontinuous fiber-reinforced thermoplast
DiCoFRTS	Discontinuous fiber-reinforced thermoset
FEM	Finite element method
FFT	Fast Fourier transformation
FODF	Fiber orientation distribution function
FOT	Fiber orientation tensor
FRP	Fiber-reinforced polymer
FRTP	Fiber-reinforced thermoplast

FRTS	Fiber-reinforced thermoset
IDD	Interaction direct derivative
RVE	Representative volume element
SIP	Single inclusion problem
SMC	Sheet molding compound
UPPH	Unsaturated polyester polyurethane hybrid resin
vol.%	Volume fraction in %
wt.%	Weight fraction in %

Latin letters

$a, b, A, B, \mathcal{D}, \ldots$	Scalar quantities
$\boldsymbol{a}, \boldsymbol{b}, \boldsymbol{c}, \ldots$	First-order tensors
$\boldsymbol{A}, \boldsymbol{B}, \boldsymbol{C}, \ldots$	Second-order tensors
$\mathbb{A}, \mathbb{B}, \mathbb{C}, \ldots$	Fourth-order tensors
$\mathbb{A}_{\langle n \rangle}, \mathbb{B}_{\langle n \rangle}, \mathbb{C}_{\langle n \rangle}, \ldots$	n^{th}-order tensors, where n is larger than four
$\underline{a}, \underline{b}, \underline{c}$	Column vectors
$\underline{\underline{A}}, \underline{\underline{B}}, \underline{\underline{C}}$	Matrices
$\boldsymbol{I}$	Second-order identity tensor
$\mathbb{I}^{\mathrm{s}}$	Symmetric fourth-order identity tensor
$\boldsymbol{Q}$	Rotation tensor
$\mathbb{P}_1$	First isotropic fourth order projector: $\mathbb{P}_1 = \frac{1}{3} \boldsymbol{I} \otimes \boldsymbol{I}$
$\mathbb{P}_2$	Second isotropic fourth order projector: $\mathbb{P}_2 = \mathbb{I}^{\mathrm{s}} - \mathbb{P}_1$
$\boldsymbol{x}$	Position Vector
v	Volume
c_{F}	Fiber volume fraction
c_{M}	Matrix volume fraction
$f(\boldsymbol{n})$	Fiber orientation distribution function
c_{β}	Weight or volume fraction
$\boldsymbol{N}, \mathbb{N}, \mathbb{N}_{\langle n \rangle}$	Fiber orientation tensors of the first kind

$\boldsymbol{F}, \mathbb{F}, \mathbb{F}_{\langle n \rangle}$	Fiber orientation tensors of the second kind
$\boldsymbol{D}', \mathbb{D}', \mathbb{D}'_{\langle n \rangle}$	Fiber orientation tensors of the third kind
$\mathbb{A}$	Strain localization tensor
$\mathbb{B}$	Stress localization tensor
$\mathbb{C}$	Stiffness tensor
$\mathbb{P}_0$	Hill's polarization tensor in the reference orientation
d	Scalar damage variable
E	Young's modulus
$\underline{p}$	Vector of material parameters
$P_{\Gamma=0}, P_{\Gamma=1}$	Biaxial loading procedures
A^{I}	Area of interest
P_{i}	Fraction of intact interfaces
P_{s}	Survival probability of interfaces
p	Regularization parameter

Greek letters

$\boldsymbol{\sigma}$	Cauchy stress tensor
$\boldsymbol{\varepsilon}$	Infinitesimal strain tensor
$\boldsymbol{\varepsilon}^{\mathrm{I}}$	Average strain in the area interest
Γ	Loading ratio in biaxial testing
$\sigma_{\mathrm{I,eq}}$	Equivalent interface stress

Operators

$\boldsymbol{AB}$	Composition of second-order tensors
$\boldsymbol{A} = \mathbb{C}[\boldsymbol{B}]$	Linear mapping of a second-order tensor by a fourth-order tensor
$\boldsymbol{A} \cdot \boldsymbol{B}$	Scalar or inner product of two tensors $\boldsymbol{A}, \boldsymbol{B}$

$\boldsymbol{a} \times \boldsymbol{b}$	Cross product of two vectors $\boldsymbol{a}, \boldsymbol{b}$
$\boldsymbol{A} \otimes \boldsymbol{B}$	Dyadic product of two tensors $\boldsymbol{A}, \boldsymbol{B}$
$\boldsymbol{n}^{\otimes k}$	Higher order dyadic product: $\boldsymbol{n}^{\otimes k} = \boldsymbol{n} \otimes \cdots \otimes \boldsymbol{n}$
$(\cdot)^{\mathsf{T}}$	Transpose of a vector or second order tensor
$(\cdot)^{\mathsf{T_M}}$	Main transpose of a fourth order tensor ($C^{\mathsf{T_M}}_{ijkl} = C_{klij}$)
$(\cdot)^{\mathsf{T_R}}$	Right transpose of a fourth order tensor ($C^{\mathsf{T_R}}_{ijkl} = C_{jikl}$)
$(\cdot)^{\mathsf{T_L}}$	Left transpose of a fourth order tensor ($C^{\mathsf{T_l}}_{ijkl} = C_{ijlk}$)
$\mathrm{symF}(\cdot)$	Full symmetrization, i.e. invariance with regard to all permutations of all indices
$(\cdot)'$	Irreducible or deviatoric part of a tensor
$\dot{(\cdot)}$	Material time derivative
$\bar{(\cdot)}$	Effective quantity
$(\cdot)_0$	Initial or reference quantity
$(\cdot)_\mathrm{F}$	Reference to fiber
$(\cdot)_\mathrm{M}$	Reference to matrix
$\langle\cdot\rangle$	Volume average
$(\cdot)^{\mathrm{SIP}}$	Reference to single inclusion problem
$(\cdot)^{\mathrm{MT}}$	Reference to Mori-Tanaka type homogenization

Bibliography

Advani, S. G., Hsiao, K.-T., 2012. Manufacturing techniques for polymer matrix composites (PMCs), 1st Edition. Woodhead Publishing, Cambridge, Philadelphia, New Delhi.

Advani, S. G., Tucker III, C. L., 1987. The use of tensors to describe and predict fiber orientation in short fiber composites. Journal of Rheology 31 (8), 751–784.

Andrusca, L., Goanta, V., Barsanescu, P. D., 2014. Optimizing the shape and size of cruciform specimens used for biaxial tensile test. Applied Mechanics and Materials 658, 167–172.

Antoniou, A., Van Hemelrijck, D., Philippidis, T., 2010. Failure prediction for a glass/epoxy cruciform specimen under static biaxial loading. Composites Science and Technology 70 (8), 1232–1241.

Asadi, A., Miller, M., Sultana, S., Moon, R. J., Kalaitzidou, K., 2016. Introducing cellulose nanocrystals in sheet molding compounds (SMC). Composites Part A: Applied Science and Manufacturing 88, 206–215.

Bandivadekar, A., Bodek, K., Cheah, L., Evans, C., Groode, T., Heywood, J., Kasseris, E., Kromer, M., Weiss, M., 2008. On the Road in 2035 – Reducing transportation's petroleum consumption and GHG emissions. Tech. Rep. July.

Bapanapalli, S., Nguyen, B. N., 2008. Prediction of elastic properties for curved fiber polymer composites. Polymer Composites 29 (5), 544–550.

Baptiste, D., 2003. Non Linear Behavior Micromechanical Multi-Scale Modelling of Discontinuous Reinforced Composites. Materials Science Forum 426-432, 3939–3944.

Bauer, J., Priesnitz, K., Schemmann, M., Brylka, B., Böhlke, T., 2016. Parametric shape optimization of biaxial tensile specimen. Proceedings in Applied Mathematics and Mechanics 16 (1), 159–160.

Belytschko, T., Bažant, Z. P., Yul-Woong, H., Ta-Peng, C., 1986. Strain-softening materials and finite-element solutions. Computers and Structures 23 (2), 163–180.

Ben Cheikh Larbi, A., Sai, K., Sidhom, H., Baptiste, D., 2006. Constitutive Model of Micromechanical Damage to Predict Reduction in Stiffness of a Fatigued SMC Composite. Journal of Materials Engineering and Performance 15 (5), 575–580.

Benveniste, Y., 1987. A new approach to the application of Mori-Tanaka's theory in composite Materials. Mechanics of Materials 6 (2), 147–157.

Benveniste, Y., 2008. Revisiting the generalized self-consistent scheme in composites: Clarification of some aspects and a new formulation. Journal of the Mechanics and Physics of Solids 56 (10), 2984–3002.

Benveniste, Y., Dvorak, G., Chen, T., 1989. Stress fields in composites with coated inclusions. Mechanics of Materials 7 (4), 305–317.

Benveniste, Y., Dvorak, G. J., Chen, T., 1991. On diagonal and elastic symmetry of the approximate effective stiffness tensor of hetereogeneous media. J. Mech. Phys. Solids. 39, 927–946.

Bertóti, R., Böhlke, T., 2017. Flow-induced anisotropic viscosity in short FRPs. Mechanics of Advanced Materials and Modern Processes 3, 1–12.

Bertram, A., 2005. Elasticity and Plasticity of Large Deformations. Springer, Heidelberg, Germany.

Biner, S., Hu, S., 2009. Simulation of damage evolution in composites: A phase-field model. Acta Materialia 57 (7), 2088–2097.

Boehler, J. P., Demmerle, S., Koss, S., 1994. A new direct biaxial testing machine for anisotropic materials. Experimental Mechanics 34 (1), 1–9.

Böhlke, T., Brüggemann, C., 2001. Graphical representation of the generalized Hooke's law. Technische Mechanik 21 (2), 145–158.

Böhlke, T., Jöchen, K., Piat, R., Langhoff, T.-A., Tsukrov, I., 2010. Elastic properties of pyrolytic carbon with axisymmetric textures. Technische Mechanik 30 (4), 343–353.

Böhlke, T., Lobos, M., 2014. Representation of Hashin-Shtrikman bounds of cubic crystal aggregates in terms of texture coefficients with application in materials design. Acta Materialia 67, 324–334.

Broutman, L. J., 1969. Measurement of the Fiber-Polymer Matrix Interfacial Strength. In: Interfaces in Composites. American Society for Testing and Materials, York, Pa, Ch. 3, pp. 27–42.

Bruderick, M., Denton, D., Shinedling, M., 2013. Applications of carbon fiber SMC for the Dodge Viper. In: Proceedings to Automotive Composites Conference & Exhibition (ACCE). Detroit.

Brylka, B., 2017. Charakterisierung und Modellierung der Steifigkeit von langfaserverstärktem Polypropylen. Doctoral Dissertation, KIT Scientific Publishing, Schriftenreihe Kontinuumsmechanik im Maschinenbau (Editor: T. Böhlke) Nr. 10.
URL https://www.ksp.kit.edu/download/1000070061

Brylka, B., Schemmann, M., Wood, J., Böhlke, T., 2018. DMA based characterization of stiffness reduction in long fiber reinforced polypropylene. Polymer Testing 66, 296–302.

Bücheler, D., 2017. Locally Continuous-fiber Reinforced Sheet Molding Compound. Ph.D. thesis, Karlsruhe Institute of Technology.

Bücheler, D., Trauth, A., Damm, A., Böhlke, T., Henning, F., Kärger, L., Seelig, T., Weidenmann, K. A., 2017. Processing of continuous-discontinuous-fiber-reinforced thermosets. In: Proceedings to SAMPE Europe Conference 2017. Stuttgart, Germany.

Buck, F., Brylka, B., Müller, V., Müller, T., Weidenmann, K. A., Hrymak, A. N., Henning, F., Böhlke, T., 2015. Two-scale structural mechanical modeling of long fiber reinforced thermoplastics. Composites Science and Technology 117, 159–167.

Budiansky, B., 1965. On the elastic moduli of some heterogeneous materials. Journal of the Mechanics and Physics of Solids 13 (4), 223–227.

Chamekh, A., Bel Hadj Salah, H., Hambli, R., 2009. Inverse technique identification of material parameters using finite element and neural network computation. The International Journal of Advanced Manufacturing Technology 44 (1-2), 173–179.

Cherif, C., 2016. Textile Materials for Lightweight Constructions. Springer Berlin Heidelberg, Berlin, Heidelberg.

Christensen, R., Lo, K., 1979. Solutions for effective shear properties in three phase sphere and cylinder models. Journal of the Mechanics and Physics of Solids 27 (4), 315–330.

Coleman, B. D., Noll, W., 1963. The thermodynamics of elastic materials with heat conduction and viscosity. Archive for Rational Mechanics and Analysis 13 (1), 167–178.

Cooreman, S., Lecompte, D., Sol, H., Vantomme, J., Debruyne, D., 2007. Identification of mechanical material behavior through inverse modeling and DIC. Experimental Mechanics 48 (4), 421–433.

Dano, M.-L., 2006. Experimental Characterization of Damage in Random Short Glass Fiber Reinforced Composites. Journal of Thermoplastic Composite Materials 19 (1), 79–96.

Davis, B. A., Gramann, P., Osswald, T., Rios, A., 2003. Compression Molding. Vol. 1. Hanser Publishers, Munich, Germany.

Demmerle, S., Boehler, J. P., 1993. Optimal design of biaxial tensile cruciform specimens. Journal of the Mechanics and Physics of Solids 41 (1), 143–181.

Deng, N., Kuwabara, T., Korkolis, Y. P., 2015. Cruciform specimen design and verification for constitutive identification of anisotropic sheets. Experimental Mechanics 55 (6), 1005–1022.

Derrien, K., Fitoussi, J., Guo, G., Baptiste, D., 2000. Prediction of the effective damage properties and failure properties of nonlinear anisotropic discontinuous reinforced composites. Computer Methods in Applied Mechanics and Engineering 185 (2-4), 93–107.

Desarmot, G., Favre, J. P., 1991. Advances in pull-out testing and data analysis. Composites Science and Technology 42 (1-3), 151–187.

Desrumaux, F., Meraghni, F., Benzeggagh, M. L., 2000. Micromechanical modelling coupled to a reliability approach for damage evolution prediction in composite materials. Applied Composite Materials 7 (4), 231–250.

Desrumaux, F., Meraghni, F., Benzeggagh, M. L., 2001. Generalised Mori-Tanaka Scheme to Model Anisotropic Damage Using Numerical Eshelby Tensor. Journal of Composite Materials 35, 603–624.

Doghri, I., Friebel, C., 2005. Effective elasto-plastic properties of inclusion-reinforced composites. Study of shape, orientation and cyclic response. Mechanics of Materials 37 (1), 45–68.

Drugan, W. J., Willis, J. R., 1996. A micromechanics-based nonlocal constitutive equation and estimates of representative volume element size for elastic composites. Journal of the Mechanics and Physics of Solids 44 (4), 497–524.

Du, D. X., Zheng, Q. S., 2002. A further exploration of the interaction direct derivative (IDD) estimate for the effective properties of multiphase composites taking into account inclusion distribution. Acta Mechanica 157 (1-4), 61–80.

Duschlbauer, D., Pettermann, H. E., Böhm, H. J., 2003. Mori-Tanaka based evaluation of inclusion stresses in composites with nonaligned reinforcements. Scripta Materialia 48 (3), 223–228.

Ehrentraut, H., Muschik, W., 1998. On symmetric irreducible tensors in d dimensions Introduction and Motivation. International Journal for Physical and Engineering Sciences ARI 51, 149–159.

Einstein, A., 1905. Eine neue Bestimmung der Moleküldimensionen. Annalen der Physik 19, 289–306.

Ernst, H., Heinrich, F., Bräuning, R., Potyra, T., Matos, E., Stadtfeld, H. C., Walch, M., 2006. Herstellungsprozess des SMC-Kofferraumdeckels des neuen VW-Cabriolets EOS und neue Technologien zur Herstellung von SMC-Bauteilen im Direktverfahren. In: 9. Internationale AVK-Tagung für verstärkte Kunststoffe und technische Duroplaste: Wettbewerbsfähig durch Verbundwerkstoffe. Essen.

Escárpita, A., Elizalde, H., Ramírez, R. A., Ledezma, E., Pinho, S. T., 2009. Modified Cruciform Specimen for Biaxial Testing of Fibre-Reinforced Composites. In: Proceedings of 17th International Conference on Composite Materials. Edinburgh, Scotland.

Eshelby, J. D., 1957. The determination of the elastic field of an ellipsoidal inclusion, and related problems. Proceedings of the Royal Society of London A: Mathematical, Physical and Engineering Sciences 241 (1226), 376–396.

Eshelby, J. D., 1959. The elastic field outside an ellipsoidal inclusion. Proceedings of the Royal Society of London A: Mathematical, Physical and Engineering Sciences 252 (1271), 561–569.

Favre, J. P., Jacques, D., 1990. Stress transfer by shear in carbon fibre model composites - Part 1 Results of single-fibre fragmentation tests with thermosetting resins. Journal of Materials Science 25 (2), 1373–1380.

Ferrari, M., 1991. Asymmetry and the high concentration limit of the Mori-Tanaka effective medium theory. Mechanics of Materials 11, 251–256.

Fish, J., Filonova, V., Fafalis, D., 2015. Computational continua revisited. International Journal for Numerical Methods in Engineering 102 (3-4), 332–378.

Fish, J., Kuznetsov, S., 2010. Computational continua. International Journal for Numerical Methods in Engineering 84 (7), 774–802.

Fitoussi, J., Bocquet, M., Meraghni, F., 2013. Effect of the matrix behavior on the damage of ethylene-propylene glass fiber reinforced composite subjected to high strain rate tension. Composites Part B: Engineering 45 (1), 1181–1191.

Fitoussi, J., Bourgeois, N., Guo, G., Baptiste, D., 1996a. Prediction of the anisotropic damaged behavior of composite materials: introduction of multilocal failure criteria in a micro-macro relationship. Computational Materials Science 5, 87–100.

Fitoussi, J., Guo, G., Baptiste, D., 1996b. Determination of a tridimensional failure criterion at the fibre/matrix interface of an organic-matrix/discontinuous-reinforcement composite. Composites Science and Technology 56 (7), 755–760.

Fitoussi, J., Guo, G., Baptiste, D., 1998. A statistical micromechanical model of anisotropic damage for S.M.C. composites. Composites Science and Technology 58 (5), 759–763.

Fitoussi, J., Meraghni, F., Jendli, Z., Hug, G., Baptiste, D., 2005. Experimental methodology for high strain-rates tensile behaviour analysis of polymer matrix composites. Composites Science and Technology 65 (14), 2174–2188.

Forest, S., Lorentz, E., 2004. Chapter XI: Localization Phenomena and Regularization Methods. In: Local approach to fracture. pp. 311–372.

Forte, S., Vianello, M., 1996. Symmetry classes for elasticity tensors. Journal of Elasticity 43 (2), 81–108.

Gower, M. R. L., Shaw, R. M., 2010. Towards a planar cruciform specimen for biaxial characterisation of polymer matrix composites. Applied Mechanics and Materials 24-25, 115–120.

Green, D. E., Neale, K. W., MacEwen, S. R., Makinde, A., Perrin, R., aug 2004. Experimental investigation of the biaxial behaviour of an aluminum sheet. International Journal of Plasticity 20 (8-9), 1677–1706.

Gross, D., Seelig, T., 2011. Bruchmechanik: Mit einer Einführung in die Mikromechanik. Springer, Heidelberg, Dordrecht, London, New York.

Guo, G., Fitoussi, J., Baptiste, D., 1997. Modelling of damage behavior of a short-fiber reinforced composite structure by the finite element analysis using a micro-macro law. International Journal of Damage Mechanics 6, 278–299.

Gutiérrez, J. C., Lozano, A., Manzano, A., Flores, M. S., 2016. Numerical and experimental analysis for shape improvement of a cruciform composite laminates specimen. Fibres and Textiles in Eastern Europe 24 (2), 89–94.

Hangs, B., Bücheler, D., Karcher, M., Henning, F., 2016. High-volume production of structural automobile parts: Comparative study of relevant composite technologies. In: FIPCO Functional Integrated Plastic Components.

Hannon, A., Tiernan, P., 2008. A review of planar biaxial tensile test systems for sheet metal. Journal of Materials Processing Technology 198 (1-3), 1–13.

Hartmann, S., Gilbert, R. R., 2017. Identifiability of material parameters in solid mechanics. Archive of Applied Mechanics 1, 1–24.

Hashin, Z., 1968. Assessment of the Self Consistent Scheme Approximation: Conductivity of Particulate Composites. Journal of Composite Materials 2 (3), 284–300.

Hashin, Z., Shtrikman, S., 1962. A variational approach to the theory of the elastic behaviour of polycrystals. Journal of the Mechanics and Physics of Solids 10 (4), 343–352.

Haupt, P., 2002. Continuum Mechanics and Theory of Materials, 2nd Edition. Springer-Verlag, Berlin, Heidelberg, New York.

Henning, F., Moeller, E., 2011. Verarbeitung faserverstärkter Kunststoffe. In: Handbuch Leichtbau. Carl Hanser Verlag GmbH & Co. KG, Munich, pp. 603–668.

Heuss, R., Müller, N., van Sintern, W., Starke, A., Tschiesner, A., 2012. Lightweight, heavy impact - How carbon fiber and other lightweight materials will develop across industries and specifically in automotive. Tech. rep., McKinsey&Company.

Hill, R., 1963. Elastic properties of reinforced solids: some theoretical principals. Journal of the Mechanics and Physics of Solids 11 (5), 357–372.

Hill, R., 1965a. A self-consistent mechanics of composite materials. Journal of the Mechanics and Physics of Solids 13 (4), 213–222.

Hill, R., 1965b. Continuum micro-mechanics of elastoplastic polycrystals. Journal of the Mechanics and Physics of Solids 13, 89–101.

Hoferlin, E., Van Bael, A., Van Houtte, P., Steyaert, G., De Maré, C., 2000. Design of a biaxial tensile test and its use for the validation of crystallographic yield loci. Modelling and Simulation in Materials Science and Engineering 8 (4), 423–433.

Hohberg, M., Kärger, L., Bücheler, D., Henning, F., 2017a. Rheological In-Mold Measurements and Characterizations of Sheet-Molding-Compound (SMC) Formulations with Different Constitution Properties by Using a Compressible Shell Model. International Polymer Processing 32 (5), 659–668.

Hohberg, M., Kärger, L., Henning, F., Hrymak, A., 2017b. Rheological measurements and rheological shell model Considering the compressible behavior of long fiber reinforced sheet molding compound (SMC). Composites Part A: Applied Science and Manufacturing 95, 110–117.

Holzapfel, G. A., 2000. Nonlinear Solid Mechanics: A Continuum Approach for Enineering. John Wiley & Sons, Ltd., New York.

Hotz, I., Schultz, T., 2015. Visualization and Processing of Higher Order Descriptors for Multi-Valued Data. Springer International Publishing Switzerland, Cham, Heidelberg, New York, Dordrecht, London.

Hour, K.-Y., Sehitoglu, H., 1993. Damage Development in a Short Fiber Reinforced Composite. Journal of Composite Materials 27 (8), 782–805.

ISO 16842, 2014. Metallic materials – Sheet and strip – Biaxial tensile testing method using a cruciform test piece. International Organization for Standardization 16842.

Jansen, C., 2013. Isogrid-stiffened automotive suspension control arm. JEC Composites Magazine 6 (82), 38.

Jeffery, G. B., 1922. The motion of ellipsoidal particles immersed in a viscous fluid. Proceedings of the Royal Society A: Mathematical, Physical and Engineering Sciences 102, 161–179.

Jendli, Z., Fitoussi, J., Meraghni, F., Baptiste, D., 2005. Anisotropic strain rate effects on the fibre-matrix interface decohesion in sheet moulding compound composites. Composites Science and Technology 65 (3-4), 387–393.

Jendli, Z., Meraghni, F., Fitoussi, J., Baptiste, D., 2004. Micromechanical analysis of strain rate effect on damage evolution in sheet molding compound composites. Composites Part A: Applied Science and Manufacturing 35 (7-8), 779–785.

Jendli, Z., Meraghni, F., Fitoussi, J., Baptiste, D., 2009. Multi-scales modelling of dynamic behaviour for discontinuous fibre SMC composites. Composites Science and Technology 69 (1), 97–103.

Jöchen, K., 2013. Homogenization of the Linear and Non-linear Mechanical Behavior of Polycrystals. Vol. 4. KIT Scientific Publishing, Karlsruhe. URL https://www.ksp.kit.edu/9783866449718

Ju, J. W., Lee, H. K., 2000. A micromechanical damage model for effective elastoplastic behavior of ductile matrix composites considering evolutionary complete particle debonding. Computer Methods in Applied Mechanics and Engineering 183 (3-4), 201–222.

Kachanov, L., 1958. Time of the Rupture Process under Creep Conditions. Izvestiia Akademii Nauk SSSR, Otdelenie Teckhnicheskikh Nauk 8, 26–31.

Kammoun, S., Doghri, I., Brassart, L., Delannay, L., 2015. Micromechanical modeling of the progressive failure in short glass-fiber reinforced thermoplastics - First Pseudo-Grain Damage model. Composites Part A: Applied Science and Manufacturing 73, 166–175.

Kanatani, K.-I., 1984. Distribution of directional data and fabric tensors. International Journal of Engineering Science 22 (2), 149–164.

Karush, W., 1939. Minima of Functions of Several Variables with Inequalities as Side Constraints. Ph.D. thesis, University of Chicago.

Katayama, T., Shinohara, M., Hakotani, M., Kitade, A., Kono, D., 2001. Development of CAE for predicting stiffness of SMC structure estimation of young's modulus in rib part. Journal of Materials Processing Technology 119 (1–3), 237–243.

Kelly, D. A., 1976. Problems in creep testing under biaxial stress systems. Journal of Strain Analysis 11 (1), 1–6.

Kinloch, A. J., Young, R. J., 1983. Fracture Behavior of Polymers. Applied Science Publishers, Barking, Essex, England.

Kosker, R., Akbarov, S. D., 2003. Influence of the Interaction between Two Neighboring Periodically Curved Fibers on the Stress Distribution in a Composite Material. Mechanics of Composite Materials 39 (2), 165–176.

Koyanagi, J., Nakatani, H., Ogihara, S., 2012. Comparison of glass-epoxy interface strengths examined by cruciform specimen and single-fiber pull-out tests under combined stress state. Composites Part A: Applied Science and Manufacturing 43 (11), 1819–1827.

Krajcinovic, D. K., 1984. Continuum damage mechanics. Applied Mechanics Reviews 37 (1), 1–6.

Krawietz, A., 1986. Materialtheorie. Springer-Verlag, Berlin, Heidelberg, New York, Tokyo.

Kröner, E., 1977. Bounds for effective elastic moduli of disordered materials. Journal of the Mechanics and Physics of Solids 25 (3), 137–155.

Kuhn, H. W., Tucker, A. W., 1951. Nonlinear Programming. In: Berkeley Symposium on Mathematical Statistics and Probability. Scientific Research Publish, Berkeley.

Kuwabara, T., Ikeda, S., Kuroda, K., 1998. Measurement and analysis of differential work hardening in cold-rolled steel sheet under biaxial tension. Journal of Materials Processing Technology 80-81, 517–523.

Lamkanfi, E., Van Paepegem, W., Degrieck, J., Ramault, C., Makris, A., Van Hemelrijck, D., 2010. Strain distribution in cruciform specimens subjected to biaxial loading conditions. Part 2: Influence of geometrical discontinuities. Polymer Testing 29 (1), 132–138.

Lecompte, D., Smits, A., Sol, H., Vantomme, J., Van Hemelrijck, D., 2007. Mixed numerical-experimental technique for orthotropic parameter identification using biaxial tensile tests on cruciform specimens. International Journal of Solids and Structures 44 (5), 1643–1656.

Lee, H. K., Simunovic, S., 2000. Modeling of progressive damage in aligned and randomly oriented discontinuous fiber polymer matrix composites. Composites Part B: Engineering 31, 77–86.

Lee, H. K., Simunovic, S., 2001. A damage constitutive model of progressive debonding in aligned discontinuous fiber composites. International Journal of Solids and Structures 38 (5), 875–895.

Lemaitre, J., 1996. A Course on Damage Mechanics. Springer Berlin Heidelberg, Berlin, Heidelberg.

Mahnken, R., Stein, E., 1996a. A unified approach for parameter identification of inelastic material models in the frame of the finite element method. Computer Methods in Applied Mechanics and Engineering 136 (3-4), 225–258.

Mahnken, R., Stein, E., 1996b. Parameter identification for viscoplastic models based on analytical derivatives of a least-squares functional and stability investigations. International Journal of Plasticity 12 (4), 451–479.

Makinde, A., Thibodeau, L., Neale, K. W., 1992. Development of an apparatus for biaxial testing using cruciform specimens. Experimental Mechanics 32 (2), 138–144.

Makris, a., Vandenbergh, T., Ramault, C., Van Hemelrijck, D., Lamkanfi, E., Van Paepegem, W., 2010. Shape optimisation of a biaxially loaded cruciform specimen. Polymer Testing 29 (2), 216–223.

Markis, A., Ramault, C., Smits, A., Van Hemelrijck, D., Clarke, A., Williamson, C., Gower, M., Shaw, R., Mera, R., Lamkanfi, E., Van Paepegem, W., 2007. A Review Of Biaxial Test Methods For Composites. In: Proceedings of the 13th International Conference on Experimental Mechanics. Alexandroupolis, Greece.

Maxwell, J. C., 1873. A treatise on electricity and magnetism. Oxford: Clarendon Press, Oxford.

Meraghni, F., Benzeggagh, M. L., 1995. Micromechanical modelling of matrix degradation in randomly oriented discontinuous-fibre composites. Composites Science and Technology 55 (2), 171–186.

Meraghni, F., Blakeman, C. J., Benzeggagh, M. L., 1996. Effect of interfacial decohesion on stiffness reduction in a random discontinuous-fibre composite containing matrix microcracks. Composites Science and Technology 56 (5), 541–555.

Meraghni, F., Desrumaux, F., Benzeggagh, M. L., 2002. Implementation of a constitutive micromechanical model for damage analysis in glass mat reinforced composite structures. Composites Science and Technology 62 (16), 2087–2097.

Miehe, C., 1996. Multisurface thermoplasticity for single crystals at large strains in terms of eulerian vector updates. International Journal of Solids and Structures 33 (20-22), 3103–3130.

Morgan, R. J., O'Neal, J. E., 1977. The microscopic failure processes and their relation to the structure of amine-cured bisphenol-A-diglycidyl ether epoxies. Journal of Materials and Science 12, 1966–1980.

Mori, T., Tanaka, K., 1973. Averate stress in matrix and average elastic energy of materials with misfitting inclusions. Acta Metallurgica 21 (5), 571–574.

Motaghi, A., Hrymak, A. N., 2017. Microstructure characterization in direct sheet molding compound. Polymer Composites .

Moulinec, H., Suquet, P., 1994. Fast numerical method for computing the linear and nonlinear properties of composites. In: Comptes rendus de l'Académie des sciences. Série II, Mécanique, physique, chimie, astronomie. Paris.

Müller, V., 2016. Micromechanical modeling of short-fiber reinforced composites. Doctoral Dissertation, KIT Scientific Publishing, Schriftenreihe Kontinuumsmechanik im Maschinenbau (Editor: T. Böhlke) Nr. 6. URL https://publikationen.bibliothek.kit.edu/1000050760

Müller, V., Böhlke, T., 2016. Prediction of effective elastic properties of fiber reinforced composites using fiber orientation tensors. Composites Science and Technology 130, 36–45.

Müller, V., Brylka, B., Dillenberger, F., Glöckner, R., Böhlke, T., Kolling, S., 2015a. Homogenization of elastic properties of short-fiber reinforced composites based on measured microstructure data. Journal of Composite Materials 50 (3), 297–312.

Müller, V., Kabel, M., Andrä, H., Böhlke, T., 2015b. Homogenization of linear elastic properties of short-fiber reinforced composites – A comparison of mean field and voxel-based methods. International Journal of Solids and Structures 67, 56–70.

Müller, W. H., 2011. Streifzüge durch die Kontinuumstheorie. Springer-Verlag, Heidelberg Dordrecht London New York.

Mura, T., 1987. Micromechanics of Defects in Solids, second, re Edition. Martinus Nijhoff Publishers, Dordrecht, Boston, Lancaster.

Nemat-Nasser, S., Hori, M., 1999. Micromechanics : overall properties of heterogeneous materials, 2nd Edition. Elsevier, Amsterdam, New York.

Nguyen, B. N., Khaleel, M. A., 2004. A mechanistic approach to damage in short-fiber composites based on micromechanical and continuum damage mechanics descriptions. Composites Science and Technology 64 (5), 607–617.

Nocedal, J., Wright, S. J., 2006. Numerical optimization, 2nd Edition. Springer, New York.

Norris, A. N., 1989. An Examination of the Mori-Tanaka Effective Medium Approximation for Multiphase Composites. Journal of Applied Mechanics 56 (1), 83–88.

Notta-Cuvier, D., Lauro, F., Bennani, B., 2014. Modelling of progressive fibre/matrix debonding in short-fibre reinforced composites up to failure. International Journal of Solids and Structures 66, 140–150.

Ogierman, W., Kokot, G., 2017. Homogenization of inelastic composites with misaligned inclusions by using the optimal pseudo-grain discretization. International Journal of Solids and Structures 113-114, 230–240.

Ogihara, S., Koyanagi, J., 2010. Investigation of combined stress state failure criterion for glass fiber/epoxy interface by the cruciform specimen method. Composites Science and Technology 70 (1), 143–150.

Ohtake, Y., Rokugawa, S., Masumoto, H., 1999. Geometry determination of cruciform-type specimen and biaxial tensile test of C/C composites. Key Engineering Materials 164-165, 151–154.

Osswald, T. A., Tucker, C. L., 1989. Compression mold filling simulation of non-planar parts. International Polymer Processing 5 (2), 79.

Ostoja-Starzewski, M., 2002. Microstructural Randomness Versus Representative Volume Element in Thermomechanics. Journal of Applied Mechanics 69 (1), 25.

Ostoja-Starzewski, M., 2006. Material spatial randomness: From statistical to representative volume element. Probabilistic Engineering Mechanics 21 (2), 112–132.

Pan, W., Boyle, J., Ramlan, M., Dun, C., Ismail, M., Hakoda, K., 2010. Material plastic properties characterization using a generic algorithm and finite element method modelling of the plane-strain small punch test. In: STP Conference, Ostrava, Czech Republic.

Périé, J. N., Leclerc, H., Roux, S., Hild, F., 2009. Digital image correlation and biaxial test on composite material for anisotropic damage law identification. International Journal of Solids and Structures 46 (11-12), 2388–2396.

Pierard, O., Friebel, C., Doghri, I., 2004. Mean-field homogenization of multi-phase thermo-elastic composites: a general framework and its validation. Composites Science and Technology 64 (10-11), 1587–1603.

Pinter, P., 2017. Personal communication within a cooperation of GRK 2078.

Pinter, P., Dietrich, S., Bertram, B., Kehrer, L., Elsner, P., Weidenmann, K., 2018. Comparison and error estimation of 3D fibre orientation analysis of computed tomography image data for fibre reinforced composites. NDT & E International 95, 26–35.

Ponte Castañeda, P., Suquet, P., 1998. Nonlinear Composites. Advances in Applied Mechanics 34 (12), 172–195.

Ponte Castañeda, P., Willis, J. R., 1995. The effect of spatial distriubiton on the effective behavior of composite materials and cracked media. Journal of the Mechanics and Physics of Solids 43 (12), 1919–1951.

Ponte Castañeda, P., Willis, J. R., 1999. Variational second-order estimates for nonlinear composites. Proceedings of the Royal Society A: Mathematical, Physical and Engineering Sciences 455 (1985), 1799–1811.

Ponthot, J.-P., Kleinermann, J.-P., 2006. A cascade optimization methodology for automatic parameter identification and shape/process optimization in metal forming simulation. Computer Methods in Applied Mechanics and Engineering 195 (41-43), 5472–5508.

Priesnitz, K., 2016. Personal communication within a cooperation of GRK 2078.

Rabotnov, Y. N., 1968. Creep rupture. In: Proceedings to the 12. International Congress of Applied Mechanics.

Reuss, A., 1929. Berechnung der Fließgrenze von Mischkristallen auf Grund der Plastizitätsbedingung für Einkristalle . Zeitschrift für Angewandte Mathematik und Mechanik 9 (1), 49–58.

Rychlewski, J., 2000. A qualitative approach to Hooke's tensors. Part I. Archives of Mechanics 52 (4-5), 737 – 759.

Schemmann, M., Brylka, B., Gajek, S., Böhlke, T., 2015a. Parameter identification by inverse modelling of biaxial tensile tests for discontinous fiber reinforced polymers. Proceedings in Applied Mathematics and Mechanics 15 (1), 355–356.

Schemmann, M., Brylka, B., Müller, V., Kehrer, L., Böhlke, T., 2015b. Mean field homogenization of discontinous fiber reinforced polymers and parameter identification of biaxial tensile tests through inverse modeling. In: Proceedings of 20th International Conference on Composite Materials.

Schemmann, M., Gajek, S., Böhlke, T., 2018a. Biaxial tensile tests and microstructure-based inverse parameter identification of inhomogeneous SMC composites. In: Altenbach, H., Jablonski, F., Müller, W. H., Naumenko, K., Schneider, P. (Eds.), Advances in Mechanics of Materials and Structural Analysis: In Honor of Reinhold Kienzler. Vol. 80. Springer International Publishing, Cham, pp. 329–342.

Schemmann, M., Görthofer, J., Seelig, T., Hrymak, A., Böhlke, T., 2018b. Anisotropic meanfield modeling of debonding and matrix damage in SMC composites. Composites Science and Technology 161, 143–158.

Schemmann, M., Lang, J., Helfrich, A., Seelig, T., Böhlke, T., 2018c. Cruciform Specimen Design for Biaxial Tensile Testing of SMC. Journal of Composites Science 2 (1), 1–12.

Schneider, D., Schoof, E., Huang, Y., Selzer, M., Nestler, B., 2016. Phase-field modeling of crack propagation in multiphase systems. Computer Methods in Applied Mechanics and Engineering 312, 186–195.

Schnur, D., Zabaras, N., 1992. An inverse method for determining elastic material properties and a material interface. International Journal for Numerical Methods in Engineering 33, 2039–2057.

Schouten, J. A., 1924. Der Ricci-Kalkül. Vol. 10 of Grundlehren der mathematischen Wissenschaften. Springer Berlin Heidelberg, Berlin, Heidelberg.

Schulenberg, L., Seelig, T., Andrieux, F., Sun, D.-Z., 2017. An anisotropic elasto-plastic material model for injection-molded long fiber-reinforced thermoplastics accounting for local fiber orientation distributions. Journal of Composite Materials 51 (14), 2061–2078.

Seelig, T., 2008. Computational modeling of deformation mechanisms and failure in thermoplastic multilayer composites. Composites Science and Technology 68 (5), 1198–1208.

Serna Moreno, M., Martínez Vicente, J., López Cela, J., 2013. Failure strain and stress fields of a chopped glass-reinforced polyester under biaxial loading. Composite Structures 103, 27–33.

Shacham, M., 1986. Numerical solution of constrained non-linear algebraic equations. International Journal For Numerical Methods in Engineering 23, 1455–1481.

Shirinbayan, M., Fitoussi, J., Meraghni, F., Surowiec, B., Bocquet, M., Tcharkhtchi, A., 2015. High strain rate visco-damageable behavior of Advanced Sheet Molding Compound (A-SMC) under tension. Composites Part B: Engineering 82, 30–41.

Simulia Dassault Systèmes, 2014. Abaqus.

Smits, A., Van Hemelrijck, D., Philippidis, T. P., Cardon, A., 2006. Design of a cruciform specimen for biaxial testing of fibre reinforced composite laminates. Composites Science and Technology 66 (7-8), 964–975.

Spahn, J., Andrä, H., Kabel, M., Müller, R., 2014. A multiscale approach for modeling progressive damage of composite materials using fast Fourier transforms. Computer Methods in Applied Mechanics and Engineering 268, 871–883.

Suquet, P., 1987. Elements of Homogenization for Inelastic Solid Mechanics. Springer-Verlag, Berlin.

Suquet, P., 1995. Overall properties of nonlinear composites : a modified secant moduli theory and its link with Ponte Castañeda's nonlinear variational procedure. Comptes Rendus de l'Académie des Sciences 320, 563–571.

Swentek, I. N., 2014. On the Interfacial Fracture Mechanics of Long- fibre Reinforced Polymer Composites. Ph.D. thesis.

Tandon, G. P., Kim, R. Y., Bechel, V. T., 2002. Fiber-Matrix Interfacial Failure Charcterization Using a Cruciform-shaped Specimen. Journal of Composite Materials 36 (23), 2667–2691.

Thom, H., 1998. A review of the biaxial strength of fibre-reinforced plastics. Composites Part A: Applied Science and Manufacturing 29 (8), 869–886.

Torquato, S., 2002. Random Heterogeneous Materials: Microstructure and Macroscopic Properties. Springer-Verlag, New York.

Trauth, A., 2017. Personal communication within a cooperation of GRK 2078.

Trauth, A., Bondy, M., Weidenmann, K., Altenhof, W., 2018. Mechanical properties and damage evolution of a structural sheet molding compound based on a novel two step curing resin system. Materials & Design 143, 224–237.

Truesdell, C., Toupin, R. A., 1960. Encyclopedia of Physics: Principles of Thermodynamics and Statics. Springer-Verlag, Berlin, Heidelberg.

Tucker III, C. L., Liang, E., 1999. Stifness predictions for unidirectional short-fiber composites: Review and evaluation. Composites Science and Technology 59, 655–671.

Van Hemelrijck, D., Ramault, C., Markis, A., Clarke, A. R., Williamson, C., Gower, M., Shaw, R., Mera, R., Lamkanfi, E., Van Paepegem, W., 2007. Biaxial testing of fibre reinforced composites. In: Proceedings of 16th International Conference on Composite Materials. Kyoto, Japan.

Voigt, W., 1889. Über die Beziehung zwischen den beiden Elastizitätskonstanten isotropter Körper. Wiedmanns Annalen 38, 573–587.

Walpole, L., 1981. Elastic Behavior of Composite Materials: Theoretical Foundations. Advances in Applied Mechanics 21, 169–242.

Weibull, W., 1951. A Statistical Distribution Function of Wide Applicability. Journal of Applied Mechanics 18, 293–297.

Willis, J. R., 1977. Bounds and self-consistent estimates for the overall properties of anisotropic composites. Journal of the Mechanics and Physics of Solids 25 (3), 185–202.

Willis, J. R., 1981. Variational and related methods for the overall properties of composites. Advances in Applied Mechanics 21, 1–78.

Yang, W., Pan, Y., Pelegri, A. A., 2012. Multiscale modeling of matrix cracking coupled with interfacial debonding in random glass fiber composites based on volume elements. Journal of Composite Materials 47 (27), 3389–3399.

Zaïri, F., Naït-Abdelaziz, M., Gloaguen, J. M., Bouaziz, A., Lefebvre, J. M., 2008. Micromechanical modelling and simulation of chopped random fiber reinforced polymer composites with progressive debonding damage. International Journal of Solids and Structures 45 (20), 5220–5236.

Zheng, Q. S., Du, D. X., 2001. An explicit and universally applicable estimate for the effective properties of multiphase composites which accounts for inclusion distribution. Journal of the Mechanics and Physics of Solids 49, 2765–2788.

Zheng, R., Tanner, R. I., Fan, X.-J., 2011. Injection Molding. Springer Berlin Heidelberg, Berlin, Heidelberg.

Zhou, S., Hrymak, A. N., Kamal, M. R., 2017. Electrical, morphological and thermal properties of microinjection molded polyamide 6/multiwalled carbon nanotubes nanocomposites. Composites Part A: Applied Science and Manufacturing 103, 84–95.

Schriftenreihe Kontinuumsmechanik im Maschinenbau
Karlsruher Institut für Technologie (KIT)
(ISSN 2192-693X)

Herausgeber: Prof. Dr.-Ing. Thomas Böhlke

Band 1 Felix Fritzen
Microstructural modeling and computational homogenization of the physically linear and nonlinear constitutive behavior of micro-heterogeneous materials. 2011
ISBN 978-3-86644-699-1

Band 2 Rumena Tsotsova
Texturbasierte Modellierung anisotroper Fließpotentiale. 2012
ISBN 978-3-86644-764-6

Band 3 Johannes Wippler
Micromechanical finite element simulations of crack propagation in silicon nitride. 2012
ISBN 978-3-86644-818-6

Band 4 Katja Jöchen
Homogenization of the linear and non-linear mechanical behavior of polycrystals. 2013
ISBN 978-3-86644-971-8

Band 5 Stephan Wulfinghoff
Numerically Efficient Gradient Crystal Plasticity with a Grain Boundary Yield Criterion and Dislocation-based Work-Hardening. 2014
ISBN 978-3-7315-0245-6

Band 6 Viktor Müller
Micromechanical modeling of short-fiber reinforced composites. 2016
ISBN 978-3-7315-0454-2

Die Bände sind unter www.ksp.kit.edu als PDF frei verfügbar oder als Druckausgabe bestellbar.

Band 7 Florian Rieger
Work-hardening of dual-phase steel. 2016
ISBN 978-3-7315-0513-6

Band 8 Vedran Glavas
Micromechanical Modeling and Simulation of Forming Processes. 2017
ISBN 978-3-7315-0602-7

Band 9 Eric Bayerschen
Single-crystal gradient plasticity with an accumulated plastic slip: Theory and applications. 2017
ISBN 978-3-7315-0606-5

Band 10 Bartholomäus Brylka
Charakterisierung und Modellierung der Steifigkeit von langfaserverstärktem Polypropylen. 2017
ISBN 978-3-7315-0680-5

Band 11 Rudolf Neumann
Two-Scale Thermomechanical Simulation of Hot Stamping. 2017
ISBN 978-3-7315-0714-7

Band 12 Mauricio Lobos Fernández
Homogenization and materials design of mechanical properties of textured materials based on zeroth-, first- and second-order bounds of linear behavior. 2018
ISBN 978-3-7315-0770-3

Band 13 Malte Schemmann
Biaxial Characterization and Mean-field Based Damage Modeling of Sheet Molding Compound Composites. 2018
ISBN 978-3-7315-0818-2